Totaro Imasaka (Ed.)

Multi-Color Laser Emission for the Generation of Ultrashort Optical Pulse

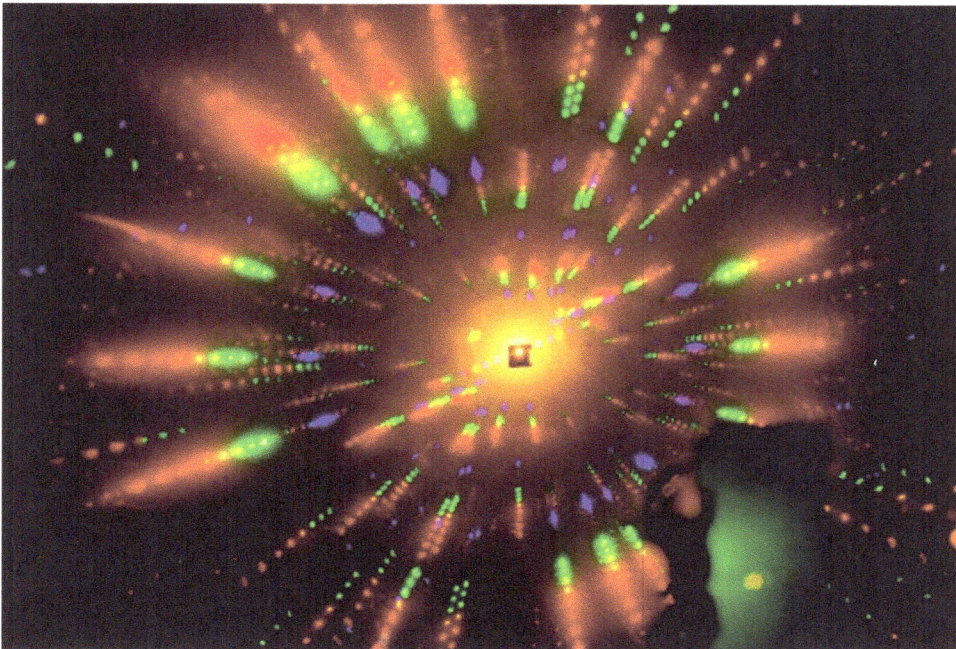

MDPI

This book is a reprint of the Special Issue that appeared in the online, open access journal, *Applied Sciences* (ISSN 2076-3417) from 2014–2016, available at:

http://www.mdpi.com/journal/applsci/special_issues/generation-ultrashort-optical-pulse

Guest Editor
Totaro Imasaka
Department of Applied Chemistry, Kyushu University
Japan

Editorial Office
MDPI AG
St. Alban-Anlage 66
Basel, Switzerland

Publisher
Shu-Kun Lin

Senior Assistant Editor
Yurong Zhang

1. Edition 2016

MDPI • Basel • Beijing • Wuhan • Barcelona • Belgrade

ISBN 978-3-03842-282-2 (Hbk)
ISBN 978-3-03842-283-9 (PDF)

Table of Contents

Section 1: Reviews

Section 2: Letters & Articles

Chapter 1: Generation of Multi-Color Laser Emission

Chapter 2: Measurement of Ultrashort Optical Pulse

Chapter 3: Spectroscopy

List of Contributors

Helder M. Crespo Departamento de Física e Astronomia e IFIMUP-IN, Faculdade de Ciências, Universidade do Porto, R. do Campo Alegre 687, Porto 4169-007, Portugal.

Chong Fang Department of Physics, and Department of Chemistry and the Center for Sustainable Materials Chemistry, Oregon State University, 153 Gilbert Hall, Corvallis, OR 97331, USA.

David C. Gold Department of Physics, University of Wisconsin at Madison, 1150 University Avenue, Madison, WI 53706, USA.

Akifumi Hamachi Department of Applied Chemistry, Graduate School of Engineering, Kyushu University, Fukuoka 819-0395, Japan.

Jinping He Japan Science and Technology Agency (JST), Core Research for Evolutional Science and Technology (CREST), 5 Sanbancho, Chiyoda-ku, Tokyo 102-0075, Japan; Advanced Ultrafast Laser Research Center, University of Electro-Communications, 1-5-1 Chofugaoka, Chofu, Tokyo 182-8585, Japan.

Xia Hua Department of Physics and Astronomy, Institute for Quantum Science and Engineering, Texas A&M University, College Station, TX 77843-4242, USA.

Tomoko Imasaka Laboratory of Chemistry, Graduate School of Design, Kyushu University, Fukuoka 815-8540, Japan.

Totaro Imasaka Department of Applied Chemistry, Graduate School of Engineering, and Division of Optoelectronics and Photonics, Center for Future Chemistry, Kyushu University, 744, Motooka, Nishi-ku, Fukuoka 819-0395, Japan.

Yuichiro Kida Department of Applied Chemistry, Graduate School of Engineering, Kyushu University, 744, Motooka, Nishi-ku, Fukuoka 819-0395, Japan.

Takayoshi Kobayashi Department of Electrophysics, National Chiao-Tung University, 1001 Ta Hsinchu Rd., Hsinchu 300, Taiwan; Japan Science and Technology Agency (JST), Core Research for Evolutional Science and Technology (CREST), 5 Sanbancho, Chiyoda-ku, Tokyo 102-0075, Japan; Institute of Laser Engineering, Osaka University, 2-6 Yamada-oka, Suita, Osaka 565-0971, Japan; Advanced Ultrafast Laser Research Center, University of Electro-Communications, 1-5-1 Chofugaoka, Chofu, Tokyo 182-8585, Japan.

A. H. Kung Institute of Atomic and Molecular Sciences, Academia Sinica, No. 1, Roosevelt Rd., Sec. 4, Taipei 10617, Taiwan; Institute of Photonics Technologies, National Tsing Hua University, Hsinchu 30013, Taiwan.

Yen-Yin Lin Brain Research Center, Department of Electrical Engineering, and Institute of Photonics Technologies, National Tsing Hua University, Hsinchu 30013, Taiwan.

Weimin Liu Department of Chemistry and the Center for Sustainable Materials Chemistry, Oregon State University, 153 Gilbert Hall, Corvallis, OR 97331, USA.

Jun Liu State Key Laboratory of High Field Laser Physics, Shanghai Institute of Optics and Fine Mechanics, Chinese Academy of Science, 390 Qinghe Rd., Jiading, Shanghai 201800, China.

Kazuya Motoyoshi Department of Applied Chemistry, Graduate School of Engineering, Kyushu University, 744 Motooka, Nishi-ku, Fukuoka 819-0395, Japan.

Tomoya Okuno Department of Applied Chemistry, Graduate School of Engineering, Kyushu University, Fukuoka 819-0395, Japan.

Jow-Tsong Shy Department of Physics, National Tsing Hua University, Hsinchu 30013, Taiwan.

Alexei V. Sokolov Department of Physics and Astronomy, Institute for Quantum Science and Engineering, Texas A&M University, College Station, TX 77843-4242, USA.

Donna Strickland Department of Physics and Astronomy, Guelph-Waterloo Physics Institute, University of Waterloo, 200 University Avenue West, Waterloo, ON N2L 3G1, Canada.

Yoshinari Takao Department of Applied Chemistry, Graduate School of Engineering, Kyushu University, 744, Motooka, Nishi-ku, Fukuoka 819-0395, Japan.

Yanli Wang Department of Chemistry and the Center for Sustainable Materials Chemistry, Oregon State University, 153 Gilbert Hall, Corvallis, OR 97331, USA.

Kai Wang Department of Physics, The University of Michigan, Ann Arbor, Michigan 48109, USA.

Joshua J. Weber Department of Physics, University of Wisconsin at Madison, 1150 University Avenue, Madison, WI 53706, USA.

Rosa Weigand Departamento de Óptica, Facultad de Ciencias Físicas, Universidad Complutense de Madrid, Avda. Complutense s/n, Madrid 28040, Spain.

Po-Shu Wu Institute of Atomic and Molecular Sciences, Academia Sinica, No. 1, Roosevelt Rd., Sec. 4, Taipei 10617, Taiwan.

Hao Yan Department of Physics and Astronomy, Guelph-Waterloo Physics Institute, University of Waterloo, 200 University Avenue West, Waterloo, ON N2L 3G1, Canada.

Hsiu-Ru Yang Institute of Photonics Technologies, National Tsing Hua University, Hsinchu 30013, Taiwan.

Deniz D. Yavuz Department of Physics, University of Wisconsin at Madison, 1150 University Avenue, Madison, WI 53706, USA.

Alexandra A. Zhdanova Department of Physics and Astronomy, Institute for Quantum Science and Engineering, Texas A&M University, College Station, TX 77843-4242, USA.

Miaochan Zhi National Institute of Standards and Technology, 100 Bureau Drive, Gaithersburg, MD 20899-8543, USA.

Liangdong Zhu Department of Physics, and Department of Chemistry and the Center for Sustainable Materials Chemistry, Oregon State University, Corvallis, OR 97331, USA.

About the Guest Editor

Totaro Imasaka, born in 1950, studied chemistry at Kyushu University. He received his doctor's degree in the field of applied chemistry in 1978. After this, he was a postdoctoral fellow at Stanford University. Since 1979 he was employed by Kyushu University. He was appointed to assistant processor, lecturer, associate professor, and full professor, in 1979, 1980, 1981, and 1991, respectively. He was promoted to distinguished professor and specially-appointed professor in 2009 and 2016, respectively. Professor Imasaka is the author or co-author of more than 310 peer-reviewed publications, 140 review/article/books, and 40 patents. He received The Chemical Society of Japan Award for Young Chemists in 1984 for his work "Study on Laser-Induced Spectrometry for Ultratrace Analysis". He also received The Divisional Award of The Chemical Society of Japan in 1994 for his piece "Studies on Two-Color Stimulated Raman Effect". More recently, he received The Award of The Japan Society for Analytical Chemistry in 2002 for his work, "Development and Application of Supersonic Jet Spectrometry and Near-Infrared Laser Spectrometry". He was a member of the editorial advisory board for the following international journals: *Chemometrics and Intelligent Laboratory Systems, Analytical Methods and Instrumentations, Fresenius' Journal of Analytical Chemistry, Talanta, Analytical Chimica Acta,* and *Analytical Chemistry.* He also served as an editor of *Analytical Sciences,* the international journal published by the Japan Society for Analytical Chemistry. He was a titular member of the International Union of Pure and Applied Chemistry (IUPAC) and is honored to serve as a fellow of this society.

Preface to "Multi-Color Laser Emission for the Generation of Ultrashort Optical Pulse"

Rainbow Stars—Multi-Color Laser Emission for Science and Art

1. Prologue: The Following Is a Nonfiction Story

December 28, 1987:

Shuichi Kawasaki, one of my graduate students, shouted for me to:

"Come to the laboratory, immediately!"

When I arrived, to my great surprise, I saw numerous laser spots twinkling very strongly on the screen.

"How fantastic! How could you generate such a colorful beam, today?" I asked him.

"I really don't know. I was trying to get better data for my thesis, and I maximized the output power of the laser which is introduced into pressurized hydrogen."

I thought this is a sort of nonlinear optical phenomenon, and that I might be able to enhance this colorful emission by increasing the output power of the dye laser. However, I immediately noticed that the dial indicating the voltage applied to the pump laser was already in the red zone, warning of possible, serious damage to the excimer laser.

"Shuichi! We might be able to slightly increase the output power by adjusting the emitting wavelength to the gain maximum, which might enhance this curious phenomenon."

"That is not correct, although I am not sure why. These colorful spots only appear when the wavelength is adjusted 6–7 nm away from the maximum of the gain curve." Shuichi said.

"I cannot believe it. This phenomenon can be observed only when the output power is at a maximum, indicating that it is generated by a type of nonlinear optical effect."

He moved the laser wavelength to the maximum of the gain curve, and surprisingly all of the colorful spots disappeared.

"Did you adjust the laser wavelength precisely at the center of the gain curve?"

He motioned with his eyes to look at the meter, indicating the laser power. I realized that the power definitely increased when the laser wavelength was changed.

"Unbelievable! Is this power meter working?" I asked.

Without saying anything, Shuichi interrupted the laser beam path. The needle went down to zero, indicating that the power meter was functioning correctly.

"Unbelievable! This cannot be possible! How can we explain the generation of this colorful beam? This appears through a nonlinear optical effect that requires a high level of power for more efficient generation. It appears, however, only when the laser power is decreased slightly from the maximum. This is logically inconsistent."

"Look! Totaro. I will change the laser wavelength and we can observe a continuous color change. This is so fantastic, and so beautiful."

All of the colorful spots, however, disappeared when the laser wavelength was moved far away from the optimum.

"Shuichi! Move the wavelength back in the opposite direction." "Oh! so fantastic!"

The spots, almost equally spaced in frequency, could be clearly seen and their colors changed continuously over the entire visible region.

"Marvelous! I can see all the rainbow colors blinking like stars!"

····Shuichi and I, two researchers with no intelligence, enjoyed ourselves by looking at the dramatic color changes until late at night····.

2. Inspiration: A Sparking of the Mechanism

When I was walking in the campus for exercise, a possible mechanism flashed through my mind: The reason why the multi-color emission is observed only when the laser wavelength is shifted 6–7 nm away from the gain maximum might be due to an amplified spontaneous emission (ASE). In the daily adjustment of the dye laser, this undesired emission, which appears at the gain maximum, needs to be reduced as much as possible. However, in the worst case, two laser emissions, i.e. an oscillating beam and ASE with a rather broad bandwidth, appear with comparable intensities. These emissions might interact with each other and produce this colorful beam, but only when the separation of the frequencies of these emissions satisfies some restriction arising from some property of the hydrogen that is used as a Raman medium. I became excited and hurried back to my office.

I asked Shuichi to measure the emission spectrum of the laser and to check the frequency separation of the oscillating beam and ASE. The value coincided exactly with the rotational Raman-shift frequency of hydrogen, verifying my above reasoning: Four-wave Raman mixing, a sort of nonlinear optical phenomenon requiring a two-color beam, played an important role in the generation of numerous rotational Raman emissions. To remove the undesirable

ASE, the manufacturer of the laser developed a new cavity design, but this, contrary to expectations, sometimes induced a strong ASE. In addition, unfortunately, or I should say fortunately, Shuichi's skill in aligning the laser was so bad (sorry) that he generated a two-color beam efficiently, but by accident. This complete story, including the defect in the laser, is described in a newsletter, published by the manufacturer of the laser.

3. Publication: Is Rainbow Stars the Name of a Toy?

I was interested in publishing this finding in a scientific journal. I nicknamed this phenomenon, "Rainbow Stars", because many spots with rainbow-colors appeared when the beam was passed through a prism and projected onto a white screen. However, the reviewer of the journal to which the paper was submitted suspected the novelty of this work and suggested removing this nickname. To publish the paper, I regretted deleting this word according to the reviewer's suggestion. A few years later, Shuichi said to me, "Do you know the term "Rainbow Stars" remained in the published paper?" I immediately checked the manuscript and noticed that we forgot to delete the term "Rainbow Stars" in the figure and the editor and reviewer didn't notice this. Shuichi and I smiled with pleasure (apologies to the editor and reviewer). Several years later, I found a toy named "Rainbow Stars" in the souvenir shop of a science museum in the USA, which was a set of colorful plastic plates containing a variety of phosphors to produce fluorescence/phosphorescence when irradiated with ultraviolet light. The name was registered as a trade mark but was issued after the publication of our paper. Because of this, I have priority in using the phrase "Rainbow Stars".

4. Art: Laser Illumination

I was unable to find any applications for this rainbow-color laser. However, it is well known that color perception is derived from three types of cone cells in the human eye, each of which has different spectral responses, which are partially superimposed on each other. Each cone cell generates and transmits a signal, at which point the color is recognized by the balance of the signals. Alternatively, all of the frequencies of light are required for the complete reproduction of a color, as is recognized from a de l'Eclairage x-y chromaticity diagram. In any case, a monochromatic multi-color laser would be desirable for the complete reproduction of a color image. It should be noted that a monochromatic light, such as a laser, easily interferes and twinkles due to the fine speckles that are induced. Such illusional light, which cannot be reproduced by photographic technology, is used in laser light shows. Because of this, such a multi-color laser might be widely used, e.g. in large-scale holography for outdoor use or in artistic presentations, entertainment, or amusement.

Due to the excellent color pictures in a magazine, I joined a society of lightening in Japan. I submitted a paper to an international journal sponsored by this society, in which I reported on the development of a high-power multi-color laser. The article contained a picture consisting of colorful beam patterns showing excellent performance of Rainbow Stars. When it was published, I was very disappointed to find that the picture was published as a "black-and-white" image and I withdrew my membership from this society a few years later.

I developed a design curriculum for undergraduate students. They made a variety of arts and devices for laser illumination. One such example was a "miracle clock" that consisted of three rotating rainbow-bars indicating the hour, minute, and second. Another interesting product was a picture of my face made of numerous optical fibers coupled with rainbow-color beams. I organized a laser show on the University campus, and many people came to enjoy this fantastic event.

5. Mistake: A Phone Call—This Is a Nonfiction Story

One day, I received a phone call.

"Hi! Totaro speaking."

"This is a phone call from the broadcast company of ⋯. I heard that you discovered a new phenomenon called Rainbow Stars. I would appreciate it if you could explain it to us."

I felt very lucky. I expected that our work would be introduced to many people via a television broadcast, and I became nervous and excited.

"That is not so great to call it a "discovery". I may have slightly exaggerated our research work when we wrote about it."

"Could you explain the phenomenon that you found in simple terms?"

I was afraid that he might not be able to understand terms such as "stimulated Raman scattering" or "four-wave Raman mixing", and even "tunable laser" or "spectroscopic measurement".

"Well. When we were focusing a laser beam into hydrogen, the laser accidentally emitted at two wavelengths⋯"

"What does "two wavelengths" mean?"

I thought that I should be more careful and not use any complicated scientific words.

"O.K. First, there were two colors, and then many colors appeared⋯"

However, it became difficult to explain our research while using no scientific words. I was confused myself and found it difficult to re-construct my logic for that explanation. At that moment, a good idea occurred me.

"O.K. You are working in a broadcast company, and you may understand that all the colors are reproduced by the superposition of three primary colors. Rainbow Stars can produce all of the colors directly and furthermore all of the emissions are monochromatic."

"What does "monochromatic" mean?"

I started losing my temper, but I replied while attempting to control myself.

"It means (I was meanwhile thinking) that such a laser beam can produce beautiful and twinkling light which can never be reproduced by current technologies such as television."

(Heavens!) I immediately noticed that I had made a big mistake in telling the truth, like a criminal talking with Detective Colombo.

"Thank you very much for the interesting discussion about your recent success. I hope to contact you for a more detailed discussion at a later time."

My mistake was to say that Rainbow Stars cannot be reproduced by a television, even though he wanted to give me a chance to introduce our work to many people by that medium.

Since then, no phone calls have been forthcoming from this company.

6. Idea: No Comments from the Reviewer

In 1992, a new idea occurred me. Rainbow Stars can be used to generate a series of emission lines spaced by a Raman shift frequency. Such equally-spaced emission lines are produced as longitudinal modes to generate mode-locked ultrashort pulses. The major difference from a conventional mode-locked laser is its wide frequency domain and large spacing between the spectral lines. When Rainbow Stars was used, we were able to generate extremely-short optical pulses at an extremely-high repetition rate. I thought this is "of course" impossible, because many researchers would have proposed such a splendid idea if it could be accomplished. However, I was really interested in knowing why this cannot be achieved. Then, I submitted a paper to a journal on this topic, although I foresaw receiving negative negative comments from the reviewer (apologies to the editor and reviewer of the journal). I expected to receive the following comments from the reviewer.

"Thank you for sending me an interesting manuscript. However, such a mode-lock laser cannot be achieved as is well-known in physics. It is my honor to explain the reason for why you are wrong. ..... I wish you great success in your research field in the future, and please accept my sincere congratulations on your efforts in this area of research."

However, I shortly received a galley proof from the journal without any editor's or reviewer's comments. I was very surprised and lost the chance to withdraw my paper. Even after its publication, I thought that such an approach would be, "of course", difficult in a real world, because we needed to generate this phenomenon using a femtosecond laser with a broad spectral bandwidth to generate a single optical pulse, although a narrow-band laser is essential for the efficient generation of stimulated Raman emissions. In fact, I could not find any paper reporting the generation of a Raman emission using a femtosecond laser.

7. Finding: Data Thrown in the Trash Box–This Is a Nonfiction Story

One day, I said to one of my undergraduate students:

"Why don't you construct a two-color dye laser and generate rainbow-color emissions by focusing the beam into hydrogen. This would be very interesting for you."

I was waiting for the results of his efforts, but he appeared to have done nothing. Then, I asked him the reason for this disappointing effort. He replied that:

"I did my work last night and confirmed the generation of rainbow-color emissions. However, I stopped my study because the spectral bandwidth of the dye laser was broad, and you told me that the spectral bandwidth must be narrow enough to generate multi-color emissions."

I was wondering about his answer and said to him:

"This phenomenon occurs only when you use a narrow-band laser because of the narrow bandwidth of the Raman gain. Your reply seems to contradict this well-known consideration and would imply that you have not carried out the experiment."

He was slightly excited and said:

"I assure you that I did my work last night. If you don't think so, please look at the data I measured yesterday!"

"Where are the data?" I asked.

He pointed at a trash box in the laboratory, and I looked for the data. I noticed that the paper of the strip chart recorder was almost at the end of a roll and he stopped his work for this reason. However, a broad dye laser spectrum was observed, as he said. I was very surprised to see the data because he said he observed a Raman emission. If these findings were correct, it should be possible to generate Raman emissions using a femtosecond laser with a broad spectral bandwidth. However, I suspected his work was flawed in some way and the following year I asked a new undergraduate student to construct two dye lasers, in which the spectral bandwidths can be independently changed to study the effect of the spectral bandwidth on the generation of Raman emissions. Surprisingly, the Raman emissions were efficiently generated, even when two broadband dye lasers were used (I felt very sorry for the former undergraduate student).

8. Fantasy: Laser Display in the Exhibition

We had no budget to continue our fundamental study, and I attempted to find applications for Rainbow Stars. On one occasion, a new technician started working in my laboratory. He had graduated from a technical high school and was 18 years old. I decided to demonstrate a multi-color laser display in an exhibition in Tokyo, which was 2 hours from my town by airplane and 8 hours by train. First, we encountered trouble in transporting a Raman cell that was pressurized at several atmospheres with hydrogen. We then negotiated with a local supplier to send a hydrogen-gas cylinder directly to the location of the exhibition in Tokyo. Later, we noticed that even vacuum pump oil cannot be transported because it contains a petroleum product. We spent more than 8 hours in preparation, e.g., the adjustments of a Nd:YAG laser and other optical components for a display. We finished all the work just before the exhibition started in the morning. It was very exciting, and we enjoyed this demonstration.

I had a chance to do a laser show in Kurume City near my town, i.e., a 2-hour drive by car. This was a combination of Rainbow Stars and a large-scale optical-fiber display with a size of several square meters. Several members of the laboratory started the installation of the equipment for generating rainbow-color emissions, starting at 5PM for a night-time show. However, we noticed that a prism that is used for beam bending was missing, which was essential for the success of a laser show. I made a phone call to a postdoctoral fellow and asked him to bring a prism immediately from the laboratory to Kurume City by driving a car in a freeway. As a result, we were able to demonstrate Rainbow Stars at mid-night and were able to return to our laboratory by 2AM.

9. Challenge: Double Rainbow!–This Is a Nonfiction Story

I had an opportunity to purchase an ultraviolet (UV) femtosecond laser, which consisted of a dye laser with an excimer amplifier. I started a research program to generate rainbow-color emissions using a femtosecond laser. Based on an agreement with the manufacturer of the laser, the user was responsible for the operation and maintenance of the laser. However, the laser stopped working within a week of its installation by an engineer, who came from a foreign country. My graduate student and I attempted to make the repairs but this proved difficult. After more than one year, we noticed that one of the optical components in the laser was poorly designed and had a problem in its structure. I pointed this out and received a newly-designed component from the manufacturer. After replacing it, we were able to immediately operate the laser. The graduate student then tried to generate Raman emissions using this UV femtosecond laser. After a few days, I asked him:

"What happened?"

"Nothing happened. I cannot even transmit the laser beam through a Raman cell."

"I think you are using borate-glass windows in the Raman cell. You should use fused-silica windows to transmit a UV beam." I said.

"I am sure that I am using fused-silica windows, and that a UV light can be transmitted through the Raman cell."

"If you are correct, please explain the reason why the UV femtosecond pulse cannot be transmitted through the cell."

"That is really puzzling to me." He said.

I was very disappointed to hear his senseless answer.

I also had an opportunity to purchase a different type of femtosecond laser, i.e., a Ti:sapphire laser that can be operated at 10 Hz in the near-infrared (NIR) region. An engineer of the manufacturer said that a daily start-up of the laser requires 8 hours. I asked another graduate student to generate Raman emissions using the femtosecond laser. After several months, the graduate student came to me and said.

"I was able to observe rainbow-color emissions, today!"

"Marvelous! Don't touch any of the components, otherwise you will lose this phenomenon when you are attempting to optimize the equipment." I said.

He immediately took a picture of the multi-color emissions. One of them consisted of a ring-shaped pattern, and he named this phenomenon, "rainbow ring". It was sometimes possible to observe another rainbow-color emission in a form of a streak. He named this, a "double rainbow". Surprisingly, or I would say, as naturally expected from theory, a rainbow-color emission was observed using a femtosecond laser. My graduate student and I were very excited by this finding and were pleased with our good luck. However, the situation changed dramatically in a few weeks. The graduate student said that the ring pattern originated from an optical Kerr effect and the streak from self-focusing in the hydrogen gas. If he is correct, how can we find Rainbow Stars?

10. Passion: Generation of Rainbow-Colored Emissions–This Is a Nonfiction Story

The graduate student using the UV femtosecond laser was enthusiastically continuing his work. Finally, he found the reason for why the UV pulse did not transmit through the Raman cell. It was due to the two-photon absorption of the high-intensity UV femtosecond pulse by the fused silica window. He then expanded the beam at the window and passed through the femtosecond beam. He immediately observed rainbow-color emissions, and I was very excited about this result. However, vibrational and rotational Raman emissions were generated simultaneously, and it was difficult to exclusively generate equally-spaced rotational lines.

Another graduate student using the Ti:sapphire laser was also concentrating his efforts on generating multi-color emissions. He shouted for me:

"Today, I was able to generate rotational Raman emissions!"

"Are you sure this time?" I said.

"Absolutely, I am sure. I observed numerous twinkling spots in line when the beam was passed through a prism and projected onto a screen. They were equally spaced in frequency and were extended from the far-UV to the NIR! I was wondering whether I should call you immediately to look for yourself or not."

"When did you find them?"

"Two o'clock", he said.

"AM or PM?"

"AM", he answered.

(I thought I was happy not to have received his phone call at such an early time.)

"How were you able to obtain this result?" I asked to him.

"Today, I chirped the laser pulse for stretching."

Later, we noticed that the laser was operating under improper conditions: the laser had a slightly narrow spectral bandwidth (longer pulse width) although the engineer from the manufacturer had suggested that we should not operate the system under such unfavorable conditions because this might cause serious damage to the laser. In fact, we frequently destroyed expensive Ti:sapphire rods in the experiment. However, such a condition was favorable for suppressing undesirable effects such as self-phase modulation and self-focusing. We also noticed that the NIR laser was useful for the exclusive generation of rotational Raman emissions.

11. Dream: Molecular-Optic Modulator

One day, I was thinking about a laser cavity. I noticed that a laser consisting of equally-spaced frequencies can transmit through the cavity and that the intensity of the light can be strongly enhanced. If this is correct, all rotational Raman emissions can be resonated in the cavity under certain conditions. This consideration suggests that Rainbow Stars, a type of nonlinear optical effect, can be generated even using a continuous-wave (CW) low-intensity laser. This idea was published in a scientific journal and was also patented. In order to increase the intensity of the light in the cavity, it is necessary to increase the finesse of the cavity using a pair of mirrors with a high reflectivity, e.g., 99.98 %, and to use a monochromatic laser beam with a narrow linewidth, e.g., 100 kHz. We solved numerous problems and were able to match the laser beam and the cavity after a 5-year effort! Eventually, we were able to generate a Stokes beam using a CW

Ti:sapphire laser. An optical beat formed by the fundamental beam and the Raman beam was measured using a laboratory-made autocorrelator. Based on current technology, it is possible to modulate a laser beam in the MHz region using an acousto-optic modulator (AOM) and in the GHz region using an electro-optic modulator (EOM). We were able to modulate the laser beam in the THz region based on molecular motion. We referred to this device as a "molecular-optic modulator (MOM)".

In order to generate a train of pulses, it is necessary to generate more than three emission lines. By carefully adjusting the laser wavelength and the cavity length, it was possible to generate the second Stokes beam, in addition to the fundamental and the first Stokes beam. A train of optical pulses, generated at 17.6 THz, was measured using the autocorrelator.

12. Obsession: Dispersion Compensation in the Cavity

The spacing among the three emission lines, which consist of the fundamental and two Stokes beams, cannot be exactly equal because the refractive index changes with changing wavelengths. For this reason, the waveform of the pulse changes at different times. In order to solve this problem, it is necessary to use a coherent process such as four-wave Raman mixing (FWRM). However, generating an anti-Stokes beam by FWRM is a difficult task, due to unavoidable positive dispersion of hydrogen in the cavity. To address this issue, we developed a dispersion-compensated optical cavity for anti-Stokes generation. A pair of mirrors with negative dispersions were used and the total dispersion in the cavity was adjusted to zero by optimizing the hydrogen pressure and by adding a noble gas with different optical properties. This was not an easy task, because the cavity length, the laser wavelength, and the pressures of the hydrogen and xenon gases must be very carefully changed in a stepwise manner and be precisely optimized. A graduate student did his best and eventually was successful. After 4 years, he found the conditions needed to generate an anti-Stokes beam; the hydrogen gas pressure should be optimized at 530 ± 5 kPa at a xenon pressure of 140 kPa!

In order to generate a train of the fastest optical pulses, it is necessary to generate a three-color beam with the largest frequency separation. When we use a three-primary-color laser, it would be possible to generate the highest repetition rate using an optical (visible) beam. Such a laser was generated using the second harmonic emission (532 nm, green) as a fundamental beam to generate the Stokes beam (683 nm, red) and the anti-Stokes beam (436 nm, blue), which supports the generation of a 125-THz pulse train. If necessary, a three-primary-color laser with better visibility could be generated using deuterium (459, 532, 632 nm).

13. Epilogue: What Are Rainbow Stars?

A wide frequency domain is essential for the generation of short optical pulses, as is known from the uncertainty principle. The multi-color laser emissions generated extend from the far-UV to the NIR and might represent an approach for a breakthrough in the generation of ultrashort pulses beyond a 1-fs barrier. In fact, a sub-cycle optical pulse can be generated by this technique, which would be impossible to achieve using current technology such as high-harmonic generation producing attosecond pulses in the extreme-UV region. Therefore, an ultrashort laser pulse generated by Rainbow Stars could have a significant impact and result in scientific breakthroughs. On the other hand, it is possible to generate ultrashort and highly-repetitive optical pulses by focusing a CW laser beam into a cavity containing a Raman medium. By phase-locking in the process of FWRM, the repetition rate and the temporal pulse profile can be stabilized by reducing the phase mismatch, thus enhancing the Raman emission. In fact, a standard of the highly-repetitive optical pulses would be obtained by this technology. A three-primary color laser would be interesting not only for use in a laser display but also for generating the fastest-repetition optical pulses. Thus, this technology has the potential for use in ultra-high-speed data communication.

Rainbow Stars were introduced more than 40 times in newspapers, magazines, on television and front covers of textbooks. There are many funny but real stories about this, but I cannot disclose all of them in written form. I hope that I will have the opportunity to discuss this with many people in the future.

Totaro Imasaka
Guest Editor

Section 1:
Reviews

Tunable Multicolored Femtosecond Pulse Generation Using Cascaded Four-Wave Mixing in Bulk Materials

Jinping He, Jun Liu and Takayoshi Kobayashi

Abstract: This paper introduces and discusses the main aspects of multicolored femtosecond pulse generation using cascaded four-wave mixing (CFWM) in transparent bulk materials. Theoretical analysis and semi-quantitative calculations, based on the phase-matching condition of the four-wave mixing process, explain the phenomena well. Experimental studies, based on our experiments, have shown the main characteristics of the multicolored pulses, namely, broadband spectra with wide tunability, high stability, short pulse duration and relatively high pulse energy. Two-dimensional multicolored array generation in various materials are also introduced and discussed.

Reprinted from *Appl. Sci.* Cite as: He, J.; Liu, J.; Kobayashi, T. Tunable Multicolored Femtosecond Pulse Generation Using Cascaded Four-Wave Mixing in Bulk Materials. *Appl. Sci.* **2014**, *4*, 444–467.

1. Introduction

Tunable, ultrashort laser pulses in different spectral ranges are powerful tools with applications in scientific research including ultrafast time-resolved spectroscopy [1–7], nonlinear microscopy [8–14] and laser micro-machining [15–18]. In the case of ultrafast time-resolved spectroscopy, which is widely used in the investigation of electronic and vibrational dynamics in molecules, the absorption peaks vary from sample to sample, and some of the molecular dynamics under investigation take place in less than 100 fs. As a result, sub-20 fs pulses with a time resolution high enough to observe real-time vibrational quantum beat and that are wavelength tunable in a wide range will play a key role. Nonlinear microscopy, such as two- or three-photon and second- or third-harmonic generation (SHG/THG) microscopy, are technologies widely used in biological research. Two-photon microscopy can be used in tissue imaging with a depth of several hundred μm [8,9], and three-photon microscopy can image to a depth of 1.4 mm [10]. SHG/THG microscopy can be used to image some biological tissues without the need for fluorescent proteins or staining with dyes, and can achieve imaging depths of several hundred μm due to its use of long excitation wavelengths [11–14]. The pump laser sources used in nonlinear microscopy have pulse widths of ~100 fs or shorter, and a visible to middle-IR spectral range [8–14]. Some spectroscopy and microscopy experiments, such as the multicolor pump-probe experiment [19], two-dimensional spectroscopy [20], or multicolor nonlinear microscopy [21–23] require ultrashort pulses with several colors.

Conventionally-used ultrashort laser sources have a spectral range of 650–950 nm (Ti:sapphire laser), 1000–1100 nm (Yb-/Nd-doped solid-state laser or fiber laser), or 1550 nm (Er-doped fiber laser). Great efforts have been made to extend the spectral range using nonlinear processes [24–40]. Optical parametric amplifier (OPA) and optical parametric oscillator (OPO) technologies are among the most successful methods for generating µJ-level pulses with spectral ranges from UV to mid-IR [32–35]. Spectrally tunable few-cycle pulses can be generated using a noncollinear optical parametric amplifier (NOPA) [36–40]. Commercial NOPA setups are available from several companies, although the price is still too high for many research groups. Pulses with broadband spectra from visible to IR (known as super continuum white light) also can be generated through filamentation in gases, bulk media, or fibers [41–45], although there are problems with the stability of the supercontinuum laser pulses [46–48].

Recently, four-wave mixing (FWM) has been studied as new method for the generation of ultrashort pulses, including few-cycle pulses, with a spectral range from deep-UV (DUV) to mid-IR [49–66]. Among these results, the multicolored laser pulses that can be generated using cascaded four-wave mixing (CFWM) in transparent bulk materials are particularly attractive, due to their ultra-broadband spectral range, large wavelength tunable range and compact configurations [54–66]. Multicolored laser pulses generated by CFWM were first shown in semiconductor lasers in the 1980s [67]. Highly efficient multicolor (>4 color) signals were generated in a nearly degenerate intracavity FWM experiment in a GaAs/GaAlAs semiconductor laser with a dye laser as the pump source for both the semiconductor laser and the FWM process. This was used as a method for quantitative determination of the third-order nonlinear optical susceptibility of the semiconductor. Eckbreth then generated multicolored light (>4) with a coherent anti-Stokes Raman scattering process in several gases, and the light was used for hydrogen-fueled scramjet applications [68]. Harris and Sokolov showed that more than 13 sidebands with a spectral range from 195 nm to 2.94 µm were generated in D_2 gas by using a Raman process [69]. In 2000, Crespo and his co-workers reported multicolored (>11 color) sideband generation using a cascaded highly nondegenerate FWM process in common glass [54]. Since then, studies have been conducted using other materials, such as sapphire plate [55,56], BBO crystal [57,58], fused silica glass [59], CaF_2 [60], BK7 glass [60], and diamond [61,62]. These studies have carefully investigated the mechanism and characteristics of multicolored laser pulses. The phase-matching condition of CFWM has also been discussed and used to explain the generation of multicolored sidebands with two

4

noncollinear pump laser pulses [60]. Our experiment has shown that more than 15 spectral upshifted sidebands and two spectral downshifted pulses can be obtained with a spectral width broader than 1.8 octaves, covering the range from UV to near-IR [55–57,59,60]. The spectra of the multicolored sidebands can also be tuned in the broadband spectral range by adjusting the cross-angle of the two pump beams or simply by replacing the nonlinear media [59,60]. The pulse duration of different sidebands can be shorter than 50 fs without any extra dispersion compensation components [55–57,59,60], and sub-20 fs pulses can be obtained when the pump pulse chirp is carefully optimized [63]. Weigand and his co-workers also tried to recombine and synthesize all of the sidebands, and found that few-cycle visible-UV pulses were feasible [64,65]. The pulse energy of the first sideband can be higher than 1 μJ, with an energy conversion efficiency of around 10% [66]. A low pump threshold for multicolored sideband generation was reported when materials with high nonlinear refractive indices, such as diamond [62] or nanoparticle-doped materials [70], were used as the nonlinear medium in the experiment. A compact experimental setup for multicolored laser pulse generation was also constructed [62]. Aside from the one-dimensional multicolored sidebands discussed above, a two-dimensional (2-D) multicolor sideband array can be generated when the pump intensity is increased in various materials such as a sapphire [55,56], diamond [62], and BBO [71,72]. Characteristically, more than 10 arrays could be generated with pump energies of several to several tens of μJ [55,56,62]. CFWM, together with beam breakup due to ellipticity of pump beams or anisotropy of nonlinear media, are thought to be the main mechanisms behind this new phenomenon [62,71]. However, simulations based on the nonlinear Schrödinger equation are still needed for the phenomenon to be fully understood.

The remainder of this paper is organized as follows. In Section 2, the theoretical analysis for multicolored pulse generation is presented. In Section 3, the characteristics of multicolored pulses are shown. The experimental setups are shown in Section 3.1. In Section 3.2, the spectral characteristics are introduced, *i.e.*, the spectral range of the sidebands, the spectral width of each sideband, and the wavelength tunability of each sideband. The characterization of pulses is described in Section 3.3. Then, the pulse energy/output power and power stability are given in Section 3.4. Multicolored pulse generation with low pump threshold is discussed in Section 3.5. In Section 4, 2-D multicolored sideband arrays are introduced and discussed. Finally, conclusions and some prospects for future research directions are given in Section 5. This article is written as a summary of recent publications reported by the authors.

2. Theoretical Analysis

2.1. FWM Process

FWM was found in the first decade of the laser epoch, and it has rapidly developed in the last twenty years. FWM is a third-order optical parametric process, in which four waves interact with each other through third-order optical nonlinearity [73]. Three waves form a nonlinear polarization at the frequency of the fourth wave during the FWM process. The wave functions of the four waves can be expressed as:

$$E_j(r,t) = A_j(r)\exp[i(k_j \bullet r - \omega_j t)](j = 1,2,3,4) \tag{2.1}$$

where, ω_j and k_j are frequencies and wave vectors of the four waves, and $A_j(r) = |A_j(r)|\exp[i\phi(r)]$ is the complex amplitude.

There are two possible roadmaps of the FWM process that satisfy the conservation of photon energies and momenta. The phase-matching condition or conservation of photon energies and momenta can be written as: (i) $\omega_4 = \omega_1 + \omega_2 + \omega_3$, $k_4 = k_1 + k_2 + k_3$; (ii) $\omega_4 + \omega_3 = \omega_1 + \omega_2$, $k_4 + k_3 = k_1 + k_2$.

The case (i) involves THG and third-order sum frequency generation. We are more interested in the FWM in case (ii), where the nonlinear polarization at frequency ω_4 can be written as:

$$P^{(3)}(\omega_4) = 3\varepsilon_0\chi_{\text{eff}}^{(3)}E_1(\omega_1)E_2(\omega_2)E_3{}^*(\omega_3)\exp[i(\Delta k \bullet r)] \tag{2.2}$$

where $\chi_{\text{eff}}^{(3)}$ is the effective third-order nonlinear optical susceptibility, and $\Delta k = k_1 + k_2 + k_3 - k_4$ is the wave vector phase-mismatching in the process. By solving the coupled-wave equations for FWM shown as follows:

$$\frac{dE_4(\omega_4)}{dr} = \frac{i\omega_4}{2\varepsilon_c cn(\omega_4)}P^{(3)}(\omega_4)\exp[-i\Delta k \bullet r] \tag{2.3}$$

we can get the optical field, $E_4(\omega_4)$, of the newly generated signal.

2.2. CFWM Process

The theoretical analysis of CFWM processes for multicolored laser pulse generation is given in [60]. The schematic of CFWM processes is shown in Figure 1a. Two vectors, k_1 and k_2, correspond to the two input beams with frequencies of ω_1 and ω_2 ($\omega_1 > \omega_2$) respectively. The mth-order anti-Stokes (spectrally blue-shifted) and Stokes (spectrally red-shifted) sidebands are marked as ASm and Sm (m = 1, 2, 3...). Figure 1b–e show the phase-matching geometries for generating the first three anti-Stokes sidebands (AS1, AS2, AS3) and the first Stokes sideband (S1). Based on these phase-matching geometries, the phase-matching condition for

6

the mth-order anti-Stokes sideband can be written as: $k_{ASm} = k_{AS(m-1)} + k^{(m)}_1 - k^{(m)}_2 \approx (m+1)k^{(1)}_1 - mk^{(1)}_2$, $\omega_{ASm} \approx (m+1)\omega_1^{(1)} - m\omega_2^{(1)}$. Since the two input beams are never single-frequency lasers, $k^{(m)}_1$ and $k^{(m)}_2$ are used instead of k_1 and k_2. The values of $\omega_1^{(m)}$, $\omega_2^{(m)}$, $|k^{(m)}_1|$, and $|k^{(m)}_2|$ may be different for every step of the m FWM processes. Similarly, with $k^{(-m)}_1$ and $k^{(-m)}_2$ used instead of k_1 and k_2, the mth-order Stokes sideband will have the following phase-matching condition: $k_{Sm} = k_{S(m-1)} + k^{(-m)}_2 - k^{(-m)}_1 \approx (m+1)k^{(-1)}_2 - mk^{(-1)}_1$, $\omega_{Sm} \approx (m+1)\omega_2^{(-1)} - m\omega_1^{(-1)}$. As the lower-order signals will participate in the generation of adjacent higher-order signals as pump pulses, this process is called CFWM.

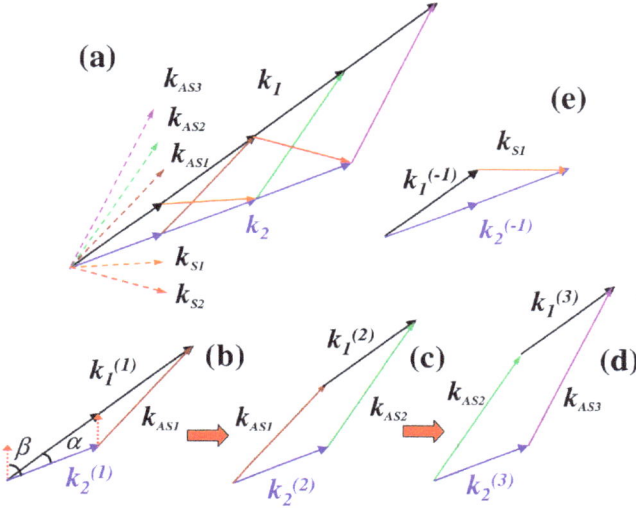

Figure 1. Schematic of multicolored sidebands generation using CFWM process. The phase-matching geometries for **(a)** AS1–AS3 and S1; **(b)** AS1; **(c)** AS2; **(d)** AS3, and **(e)** S1 [60].

Based on the phase-matching condition we have expressed above, the output parameters, such as wavelength and output angle, of the generated sidebands can be calculated to explain and inform experimental work. In our experiments, the wavelength range of the two pump beams were 660–740 nm (Beam 1) and 800 nm (Beam 2). The nonlinear medium was assumed to be fused silica plate with a thickness of 1 mm. The simulations were performed under these conditions. The wavelength dependence of Beam 1 for optimal phase-matching on the order number at different cross-angles is shown in Figure 2a. To fulfill the phase-matching condition, the wavelength of Beam 1 should redshift for higher order anti-Stokes sidebands. The wavelengths of generated sidebands for different cross-angles are shown in Figure 2b, which clearly shows that the wavelengths of same order sidebands can be tunable by changing the cross-angle of the two pump beams, and the tuning range covered the wavelength gap between adjacent sidebands. The exit angles of the generated sidebands are plotted against the order number at different cross-angles in Figure 2c. The difference in exit angle between the multicolored sidebands was large enough for easy separation, even for adjacent sidebands. The dependence of the exit angle on the center wavelength of the generated sidebands at different cross-angles in different materials is shown in Figure 2d. The phase mismatching for the first four anti-Stokes sidebands at two different angles, 1.87° and 2.34°, are shown in Figure 2e. The increase of the slope of the curves with the order numbers means the reduction of the gain bandwidth for the sidebands, which was confirmed by our experiment. The calculations based on the phase-matching condition of CFWM agree with the experimental results, which will be given in next section.

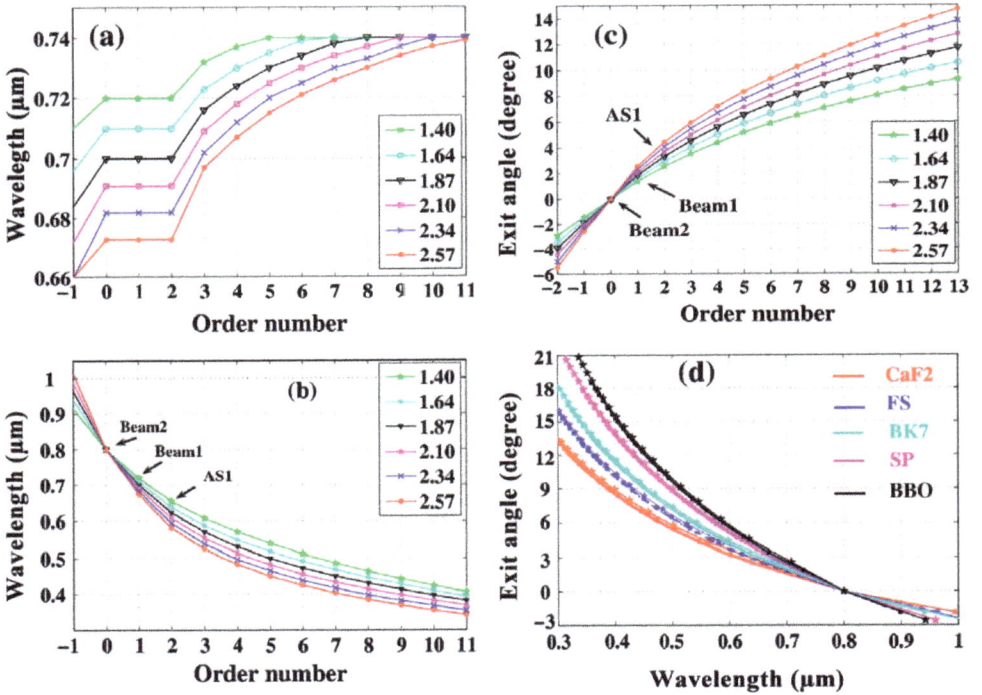

Figure 2. *Cont.*

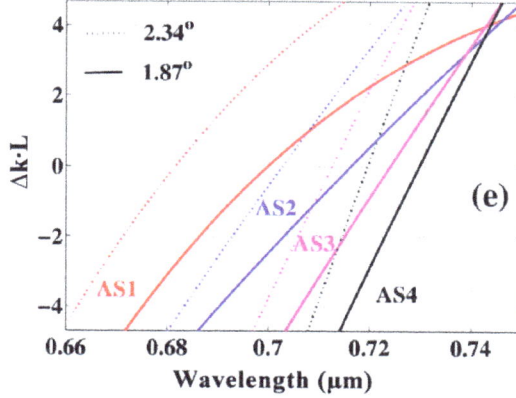

Figure 2. Calculated output parameters of generated sidebands. Dependence of (**a**) central wavelength of Beam 1 for minimum phase-mismatching; (**b**) the wavelength; and (**c**) the exit angles of generated sidebands on the order number at different crossing angles; (**d**) Dependence of exit angle of the generated sidebands on the center wavelength at three different cross-angles, 1.40°, 1.87°, and 2.57°, in different nonlinear media; (**e**) Phase mismatching of the sidebands from AS1 to AS4 at 1.87° and 2.34° in 1 mm fused silica [60].

3. Experimental Characteristics of Multicolored Pulses

3.1. Experimental Setups

Various experimental setups for multicolored laser pulse generation have been reported in the literature. The main differences between these setups are the methods for preparing the two pump laser beams. Crespo and his coworkers used two femtosecond pulses from a dye-laser amplifier system, with Beam 1 (561 nm, 40 fs) and Beam 2 (618 nm, 80 fs), and a pulse energy of 20 μJ for each beam [54]. Zhi used two OPA systems, pumped with a commercial Ti:sapphire amplifier [61]. The SHG signals of the signal and idler pulses from the two OPAs were used as the pump pulses for the generation of multicolored sidebands. The central wavelength and pulse energy of the two pump beams were 630 nm/1–3 μJ and 584 nm/1–3 μJ. There were also some other differences, including the Ti:sapphire amplifier pulse, and the supercontinuum generated in bulk materials [58].

We have used two experimental setups for multicolored pulse generation. As shown in Figure 3, we used Type-1 experimental setup for most of our work. The pump source was a 1 kHz Ti:sapphire regenerative amplifier laser system (35 fs/2.5 mJ/1 kHz/800 nm, Micra + Legend-USP, Coherent, Santa Clara, CA, USA). The pump laser was split into four beams for different uses. One beam (Beam 1), with energy of 300 μJ, was focused into a krypton-gas-filled hollow-core fiber with inner and outer diameters of 250 μm and 3 mm, and a length of 60 cm. The spectrum of Beam 1 broadened to a range extending from 600 to 950 nm after transmission through the hollow-core fiber, while the pulse energy decreased to about 190 μJ, due to coupling and propagation loss. A pair of chirped mirrors and two glass wedges were applied to compensate for the chirp of Beam 1 with broadband spectrum. A nearly transform-limited pulse, with a pulse duration of ~10 fs, was obtained by changing the bounce times on the chirped mirrors and the insertion of the glass wedges. Negatively and positively chirped pulses also can be obtained for different experiments. Beam 1 was then spectrally filtered with band-pass filters (BPF) short-wavelength-pass filters (SPF), or long-wavelength-pass filters (LPF) in different experiments. A concave mirror with a focal length of 600 cm was used to focus Beam 1 into the nonlinear medium (G1). Beam 2 was focused into the nonlinear medium by a lens with a focal length of 1 m. The fourth beam (Beam 4) was used to characterize the generated multicolored pulses with the cross-correlation frequency-resolved optical gating (XFROG) technique [74] in a 10 μm-thick BBO crystal.

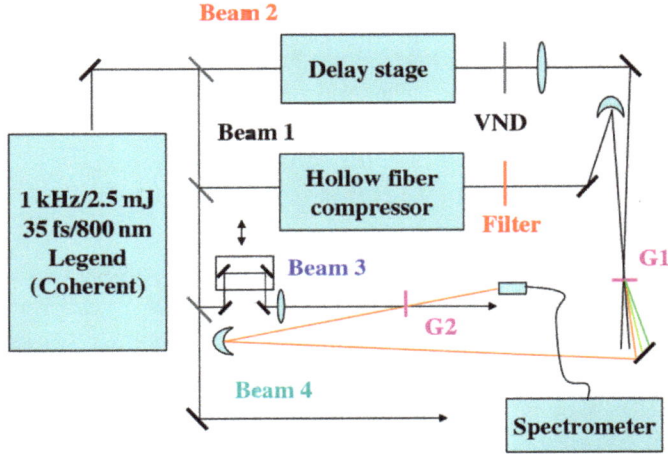

Figure 3. Type-1 experimental setup. VND: variable neutral-density filter. G1: nonlinear medium for multicolored sidebands generation. G2: nonlinear medium for pulse measurement with an X-FROG system [74].

Figure 4 shows the schematic of Type-2 setup, which was used for the generation of low-threshold multicolored sidebands and 2-D multicolored arrays in a diamond plate. Another Ti:sapphire laser system (35 fs/2.5 mJ/1 kHz/800 nm, Spitfire ACE, Spectra-Physics) was used as the pump source, and a beam with pulse energy of 150 μJ was used in the experiment. A BK7 glass plate with a thickness of 3 mm was used to spectrally broaden the laser pulse using a self-phase modulation (SPM) process. After that, a pair of chirped mirrors (GDD) was used to compensate for the dispersion induced by the BK7 glass and other components. Then, the laser beam was split into two parts, Beam 1 and Beam 2, with a beamsplitter (BS). Beam 1 first propagated through a short-pass filter (F1, cut-off wavelength of 800 nm), was then focused into the nonlinear medium by a concave mirror (M4) with a focal length of 500 mm. Beam 2 was spectrally filtered with a long-pass filter (F2, cut-on wavelength of 800 nm), and then focused into the nonlinear medium by another concave mirror (M3) with a focal length of 500 mm. The beam diameters of both Beam 1 and Beam 2 were ~300 μm on the 1 mm thick diamond plate. As no hollow fiber or gas chamber is used, the Type-2 experimental setup occupied half the space of a Type-1 setup.

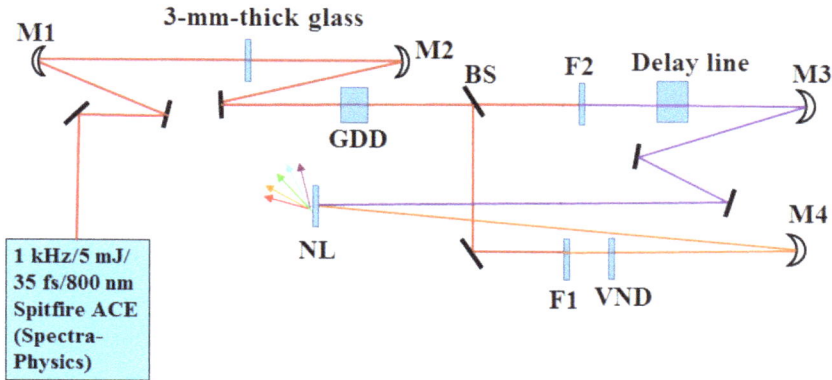

Figure 4. Type-2 experimental setup. Focal lengths of concave mirrors, M1, M3, and M4 are 500 mm, while that of M2 is 250 mm. GDD: chirped mirrors. F1: longpass filter with a cut-off wavelength of 800 nm. F2: short-pass filter with a cut-off wavelength of 800 nm. BS: beamsplitter. VND: variable neutral-density filter. NL: nonlinear medium for multicolor pulse generation.

In the experiment, the diameters of the incident beams at the position of the nonlinear media were measured using a CCD camera (BeamStar FX33, Ophir Optronics: Jerusalem, Israel). The pulses were characterized using the XFROG or SHG-FROG technique and retrieved with a commercial software package (FROG 3.0, Femtosoft Technologies: Chennai, India). The spectra of the pulses were measured with a commercial spectrometer (USB4000, Ocean Optics: Dunedin, FL, USA). To avoid optical damage, the intensities of the two pump beams on the surface of nonlinear media surface were set at least one order of magnitude lower than the damage threshold for all the media used. Neither damage nor supercontinuum generation were observed in any experiment.

3.2. Spectra and Wavelength Tuning of Multicolored Sidebands

3.2.1. Tuning the Wavelength of Sidebands by Changing Cross-Angle

The Type-1 experimental setup was applied, and a short-pass filter with cut-off wavelength of 800 nm was used to eliminate the red components of Beam 1. As a result, the spectra of the two pump pulses were as in Figure 5. The input powers of Beam 1 and Beam 2 were 11 and 19 mW, respectively.

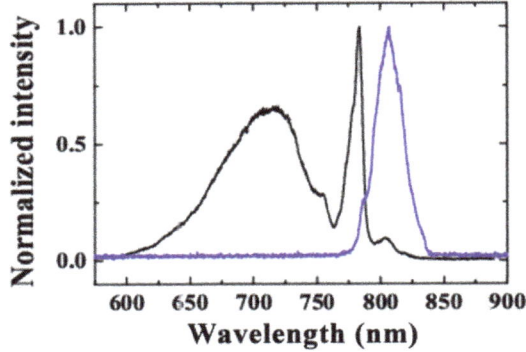

Figure 5. The spectra of the two pump beams, Beam 1 (black curve) and Beam 2 (blue curve) [75].

Multicolored sidebands were obtained when the input beams overlapped well both spatially and temporally, as shown in Figure 6a. These sidebands were in the same plane as the pump beams, but separated with different exit angles. Figure 6b depicts the spectra of the lowest four-order red-shifted sidebands, AS1-AS4, when the pump beam cross-angle was 2.1°. The spectral width decreased with an increase in the order number of the sidebands, as was predicted by the calculations, shown in Figure 2e. Figure 6c shows the central wavelength of AS2 with different pump beam cross-angles. The central wavelength of AS2 shifted from 500 nm to 625 nm, and the cross-angle increased from 1.5° to 2.5°. This tuning range successfully surpassed the wavelength gap between AS1 and AS3, as shown in Figure 6b. Therefore, it was possible to tune the wavelength continuously by simple angle tuning, without a gap between the neighboring order sidebands. The spectra of AS8-AS15 are shown in Figure 6d. The spectra of S1 and S2 are shown in Figure 6e. The whole wavelength range obtained by angle tuning of all sidebands covered the near UV-visible-near IR range from 360 nm to 1.2 μm, corresponding to more than 1.8 octaves. These broadband spectra and large tunability are very attractive and useful from the viewpoints of application and creating simple tuning mechanisms.

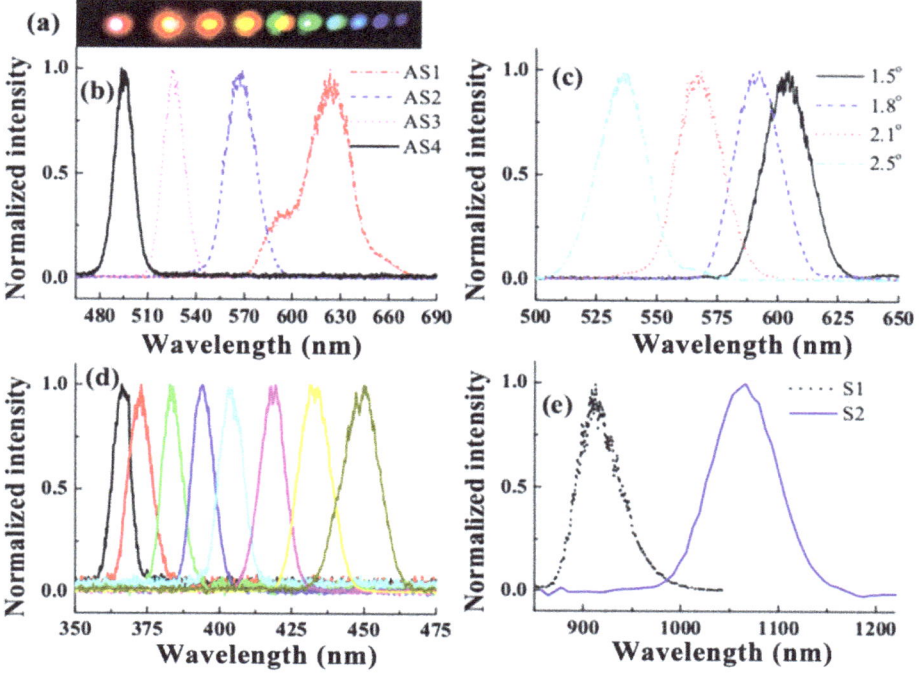

Figure 6. (**a**) Photograph of the first ten anti-Stokes sidebands on a white paper set 1 m far from the nonlinear medium; (**b**) The spectra of AS1-AS4 with cross-angle of 2.1° for two pump beams; (**c**) The spectra of AS2 with different cross-angles; Spectra of (**d**) AS8-AS15 and (**e**) two Stokes signals S1 and S2 with cross-angle of two pump beams set as 1.5° [59].

3.2.2. Tuning the Wavelength of Sidebands by Changing Nonlinear Media

The phase-matching condition for CFWM is $k_{Sm} = k_{S(m-1)} + k^{(-m)}_2 - k^{(-m)}_1 \approx (m + 1)k^{(-1)}_2 - mk^{(-1)}_1$, $\omega_{Sm} \approx (m + 1)\omega_2^{(-1)} - m\omega_1^{(-1)}$, as discussed in Section 2. The wave vectors k can be written as $k = n \times k_0$, where n is the linear refractive index of the nonlinear medium and k_0 is the wave vector in vacuum. This means that the refractive index (dispersion curve) of the medium will also influence the phase-matching conditions. The refractive index (dispersion curve) can be adjusted by replacing the medium with different optical properties. Figure 7 shows the spectra of AS1 and AS3 for nonlinear media (CaF$_2$ plate, fused silica plate, BK7 glass plate, sapphire plate, and BBO crystal) with a fixed pump beam cross-angle of 1.8°. By changing the media, the central wavelength of AS1 could be tuned from 640 nm to 610 nm, while the central wavelength of AS3 was adjusted correspondingly from 490 nm to 560 nm. The spectrum of AS3 in the BBO crystal overlapped with

13

the spectrum of AS1 in the CaF_2 crystal, which means that spectral gaps between neighboring sidebands can be bridged simply by replacing the nonlinear medium.

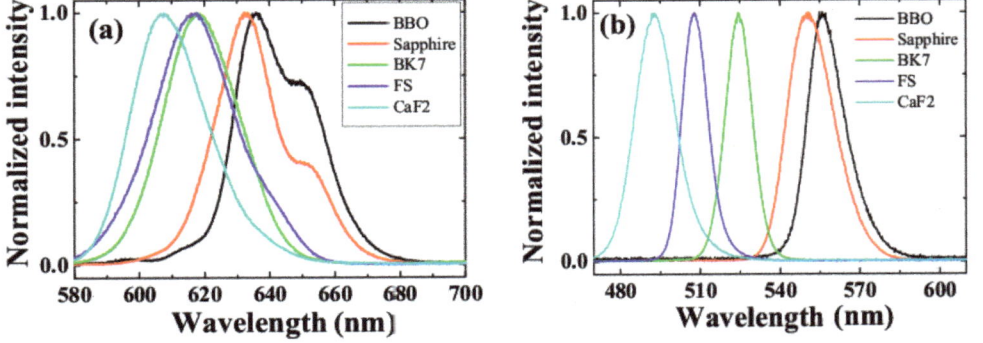

Figure 7. Spectra of (a) AS1 and (b) AS3 of five different materials with cross-angle of 1.8° [60].

3.3. Temporal Characteristics of Multicolored Pulses

The pulse durations of Beam 1 and Beam 2 were measured to be 40 fs and 55 fs, respectively. The characteristics of the sidebands, S1, AS1, and AS2, which were generated using these two pump pulses are shown in Figure 8. The pulse durations of AS1, AS2, and S1 were calculated to be 45 fs, 44 fs, and 46 fs, respectively. The retrieved phase showed that these pulses were all positively chirped, and that the positive chirp induced by material dispersion of the nonlinear medium prevented shorter pulses from being obtained.

As discussed in [63], chirped pump pulses can be used for pre-compensation of the positive chirp of the sidebands, resulting in even shorter pulse durations. The principle of the process can be explained in the following way. In the CFWM process, the m-th-order anti-Stokes sideband has the phase matching condition: $k_{ASm} = k_{AS(m-1)} + k_1 - k_2 = (m+1)k_1 - mk_2$, $\omega_{ASm} \approx (m+1)\omega_1 - m\omega_2$. The m-th ($m > 0$) order anti-Stokes signal can be expressed as follows, if the electric field of two incident pulses are given as: $E_j(t) \propto \exp\{i[\omega_{j0}t + \phi_j(t)]\}$, $j = 1, 2$

$$E_{ASm}(t) \propto \exp\{i[((m+1)\omega_{10} - m\omega_{20})t + ((m+1)\phi_1(t) - m\phi_2(t))]\} \qquad (3.1)$$

If Beam 1 is negatively chirped ($\partial^2\phi_1(t)/\partial t^2 < 0$) and Beam 2 is positively chirped ($\partial^2\phi_2(t)/\partial t^2 > 0$), we can obtain:

$$\partial^2\phi_{ASm}(t)/\partial t^2 = (m+1)\partial^2\phi_1(t)/\partial t^2 - m\partial^2\phi_2(t)/\partial t^2 < 0 \qquad (3.2)$$

14

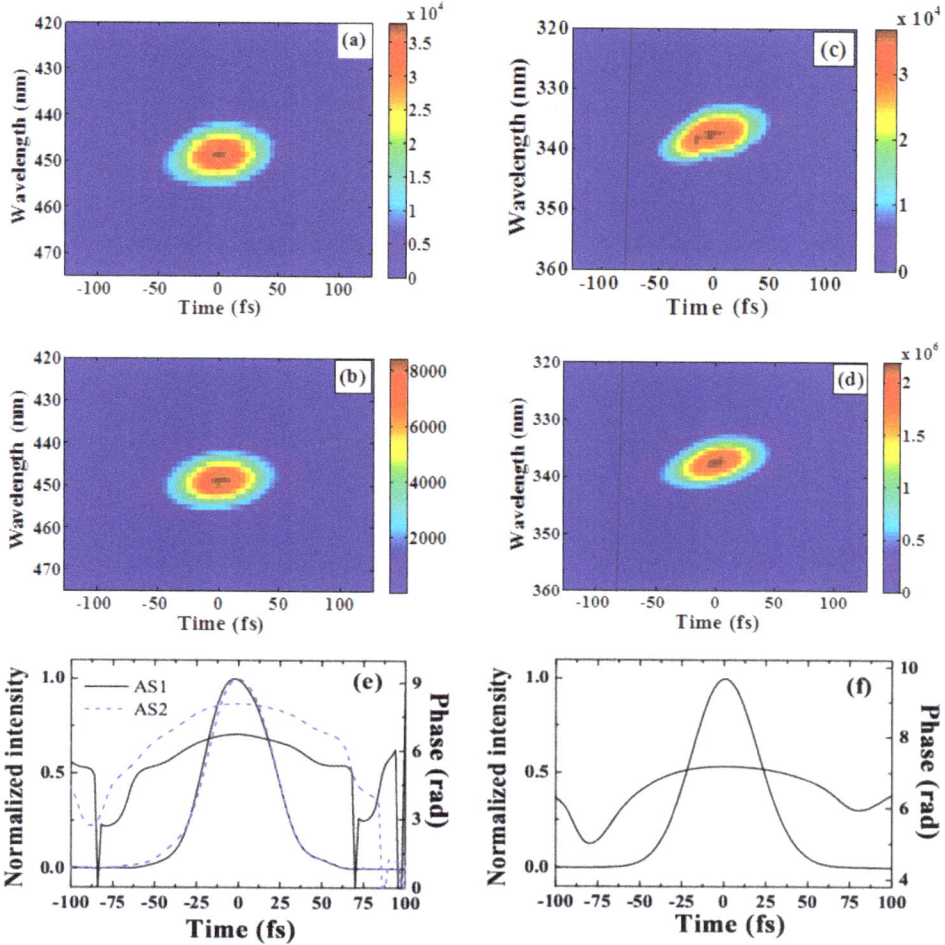

Figure 8. (a) Measured and (b) retrieved XFROG traces of S1; (c) measured and (d) retrieved XFROG traces of AS2 when the cross-angle was 1.87°. Retrieved temporal intensity profiles and phases of (e) AS1 (solid line), AS2 (dashed line) and (f) S1 [66].

This means that the m-th order blued-shifted field, E_{ASm}, can be negatively chirped. As such, a nearly transform-limited pulse can be achieved, if the negative chirp of the ASm field is precisely adjusted to correctly compensate for the dispersion induced by the nonlinear media and other optical components used in the processes of pulse generation and characterization. By this method, the pulse durations of AS1 and AS2 were compressed to 15 fs and 16 fs, respectively, as shown in Figure 9. Further optimization of the dispersion, including higher-order dispersion, is needed to obtain truly transform-limited pulses.

3.4. Output Power/Energy of Multicolored Pulses

Table 1 shows the average power of AS1-AS3 obtained with five different bulk media. The external cross-angle of the two pump beams was 1.8°, while the input powers of Beam 1 and Beam 2 were set at 6.5 and 25 mW respectively. CaF_2 had the highest AS1 output power, and the lowest AS3 output power of all five media. Conversely, in BBO crystal, the powers of the sidebands decreased the most rapidly with increasing order number. This phenomenon can be explained by the different phase-matching conditions and dispersion properties of the five materials.

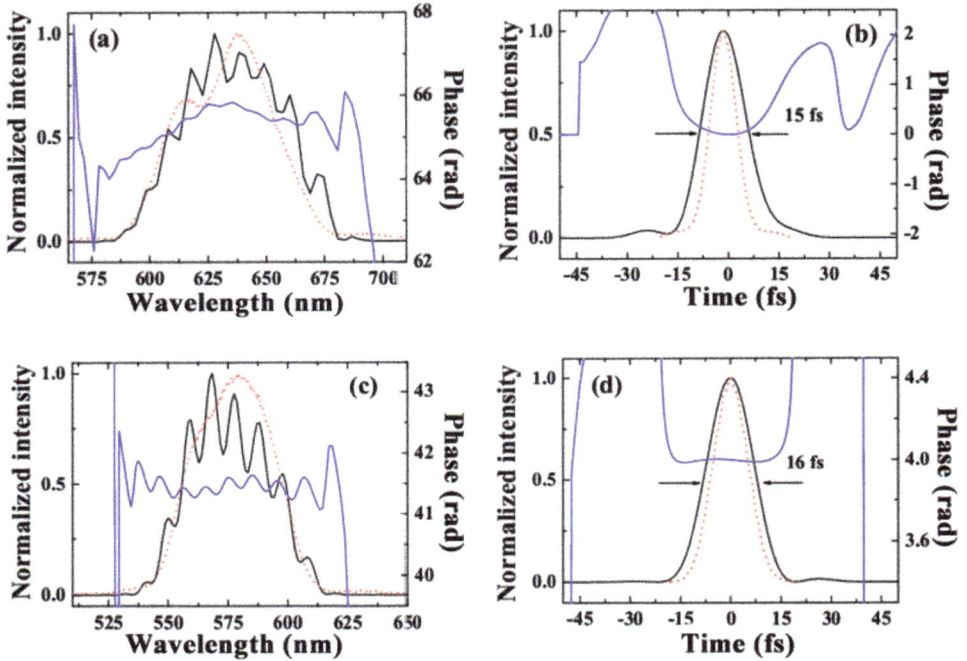

Figure 9. (a) Measured (red), retrieved (black) spectral intensity profile and retrieved spectral phase (blue) of AS1; (b) Retrieved temporal intensity profile (black), temporal phase (blue), and calculated transform-limited temporal intensity profile (red) of AS1; (c) Measured (red), retrieved (black) spectral intensities and retrieved spectral phase (blue) of AS2; (d) Retrieved temporal intensity profile (black), phase (blue), and calculated transform-limited temporal intensity profile (red) of AS2 [63].

16

Table 1. The output power of AS1-AS3 with five commonly used third-order nonlinear media. The external cross-angle of two pump beams is 1.8°, while the input powers of Beam 1 and Beam 2 are 7 mW and 25 mW, respectively [60].

μW	CaF$_2$	Fused Silica	BK7	Sapphire Plate	BBO
AS1	480	700	715	750	780
AS2	210	315	295	210	135
AS3	125	90	60	40	10

Figure 10a shows the power dependence of AS1 on the power of Beam 1, with the power of Beam 2 fixed at 19 mW and the cross-angle set at 1.8°. The output power of AS1 was sensitive to the pump power with a low pump rate, and saturation occurred when the power of Beam 1 increased to about 11 mW. Similarly, when the power of Beam 1 was set to 11 mW and Beam 2 had a high pump power, saturation of the output power of AS1 appeared, as shown in Figure 10b. This saturation may have helped to obtain sidebands with high stability. The power stability of AS1 and Beam 1 were 0.95% and 0.62% in RMS, respectively, as shown in the inset of Figure 10a.

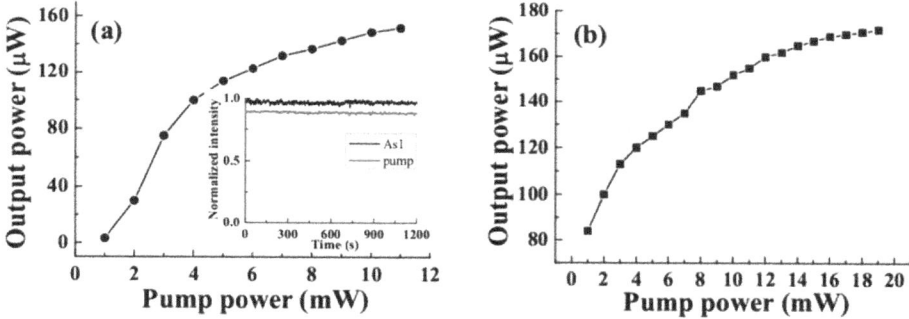

Figure 10. The power dependence of AS1 on (a) Beam 1 with power of Beam 2 fixed at 19 mW, and (b) Beam 2 with power of Beam 1 fixed at 11 mW. Power stabilities of AS1 and Beam 1 in twenty minutes are shown as the insertion of (a) [59].

By optimizing the spatial and temporal overlap of the two pump beams, the maximum pulse energy of S1 and AS1 reached was higher than 1 μJ [66]. An even higher output power was achieved by enlarging the pump beam size and increasing the pump power.

The polarization at frequency ω_{AS1} in FWM process generating AS1 can be written as:

$$P^{(3)}(\omega_{AS1}) \propto \chi_{eff}^{(3)} E^2(\omega_1) E^*(\omega_2) \tag{3.3}$$

According to the coupled-wave equations in the FWM process, the optical field and polarization at frequency ω_{AS1} had the following relationship:

$$\frac{dE(\omega_{AS1})}{dz} = \frac{i\omega_{AS1}}{2\varepsilon_0 c n_{AS1}} P^{(3)}(\omega_{AS1}) e^{-i\Delta kz} \tag{3.4}$$

Here, Δk is the wave vector mismatch in the FWM process. From Equations (3.3) and (3.4), it can be seen that the intensity of AS1 becomes higher, following the proportionality relation with the squared absolute value of the nonlinear optical susceptibility ($\left|\chi_{eff}^{(3)}\right|^2$) of the material used.

Based on this, diamond, the nonlinear optical susceptibility of which is ~5 time larger than that of sapphire and ~10 times larger than that of CaF$_2$ [73], was used in the experiment to obtain multicolored sidebands with a low threshold.

The experiment was performed with Type-2 setup shown in Section 3.1. The spectra of two pump beams, Beam 1 and Beam 2, are depicted in Figure 11. The two spectral positions were adjusted by tuning the angle between input beams and filters. The retrieved temporal intensity profiles and phases of two pump pulses are shown in Figure 12, where the Beam 1 and Beam 2 pulse durations are 81 fs, and 47 fs, respectively.

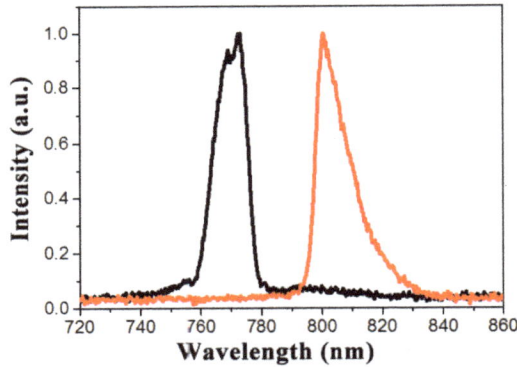

Figure 11. The spectra of Beam 1 (black), Beam 2 (red) [62].

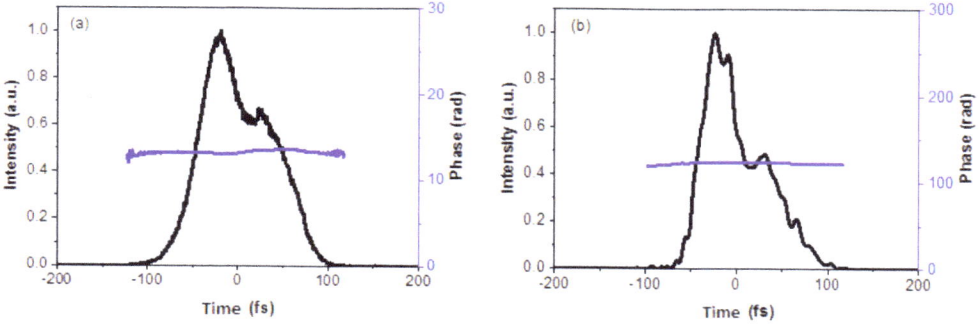

Figure 12. The retrieved intensity profile and phase of (**a**) Beam 1; (**b**) Beam 2 [62].

The average power of Beam 1 was set to 0.855 mW, and the average power of Beam 2 was continuously changed by a VND. Figure 13 shows the multicolored sidebands at different pump levels. The intensities of Beam 1 and Beam 2 on the diamond plate in Figure 13a were calculated to be 14.9×10^9 W/cm^2 and 12.3×10^9 W/cm^2 respectively. These were much lower than the threshold intensities obtained previously for multicolored sideband generation in a fused silica plate, of 60×10^9 W/cm^2 and 8×10^9 W/cm^2 [70]. This low pump threshold for multicolored sideband generation is important in the context of an actual experiment, because pump lasers with high repetition frequencies, *i.e.*, several hundred kHz to several MHz, inevitably have a low pulse energy when used with a conventional amplifier system. Compared to the multicolored sidebands generated with a 1 kHz amplifier, pulses with a higher repetition frequency are more useful in nonlinear microscopy. Low repetition frequencies make the image frame times unconventionally long and also lead to high noise levels. Figure 13b–d show the 2D structure obtained by increasing the power of Beam 2, a detailed discussion of which will be given in the next section.

Figure 14 shows the spectra of generated multicolored sidebands obtained under pump rates of 0.855 mW and 0.856 mW for Beam 1 and Beam 2, respectively. The normalized spectra, AS1-AS5, of the two pump beams, S1 and S2, are shown in Figure 14. The spectral width of these sidebands was broader than 10 nm, which means that a pulse duration of <100 fs was achievable. The spectra of these sidebands could also be continuously tuned by adjusting the cross-angle of Beam 1 and Beam 2.

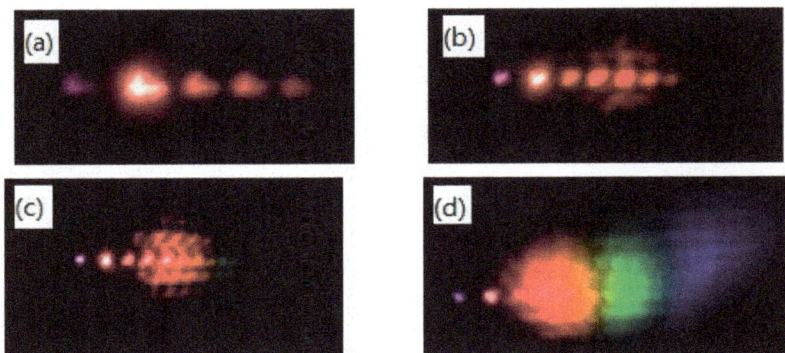

Figure 13. Multicolored pattern generated with different pump levels. (a) Beam 1: 0.855 mW, Beam 2: 0.410 mW; (b) Beam 1: 0.855 mW, Beam 2: 0.856 mW; (c) Beam 1: 0.855 mW, Beam 2: 1.121 mW; (d) Beam 1: 0.855 mW, Beam 2: 1.970 mW [62].

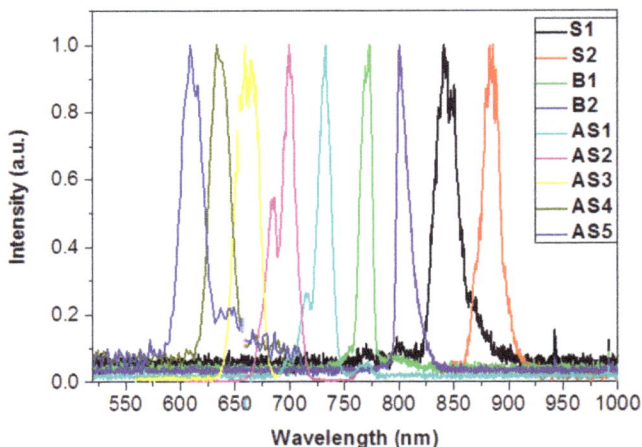

Figure 14. The normalized spectra of two pump beams, Beam 1 and Beam 2, and several sidebands AS1-AS5, S1, S2 [62].

Table 2 shows the output power of AS1-AS5 and S1 when the average power of Beam 1 and Beam 2 were set at 0.855 mW and 0.856 mW. The conversion efficiency was about 1.84%, 1.99%, 0.36%, 0.15%, 0.08% and 0.05% for S1, AS1, AS2, AS3, AS4 and AS5 respectively. The power of AS1–AS4 could be increased by increasing the pump rate, as shown in Figure 13c,d.

Table 2. Output power of AS1-AS5 and S1 with average power of 0.855 mW and 0.856 mW for Beam 1 and Beam 2, respectively [62].

Sidebands	AS1	AS2	AS3	AS4	AS5	S1
Power (μW)	34.0	6.1	2.5	1.3	0.8	31.4

4. 2-D Multicolored Sidebands Arrays

Zeng and his coworkers first observed 2-D multicolored arrays in 2006 in a quadratic nonlinear medium (BBO crystal) with two closely-overlapped femtosecond laser beams from Ti:sapphire amplifier and its SHG signal [71]. The cause of the 2-D pattern was thought to be the cascaded quadratic nonlinear process, together with spatial breakup of the quadratic spatial solitons induced by ellipticity of the input beams. The 2-D structure could also be suppressed by another weak SHG beam.

Zhi also generated 2-D multicolored arrays in diamond plate with three pump beams [61], attributed to the interaction of two different sets of cascaded stimulated Raman scattering processes.

We observed a similar structure in a cubic nonlinear medium, sapphire plate, with only two pump beams in 2008 [55,56]. 2-D multicolored arrays were generated when pump energy was increased. Figure 15 shows the 2-D multicolored arrays generated under various conditions. The 2-D multicolored arrays could be controlled by changing the intensity, delay, or polarization of one input beam.

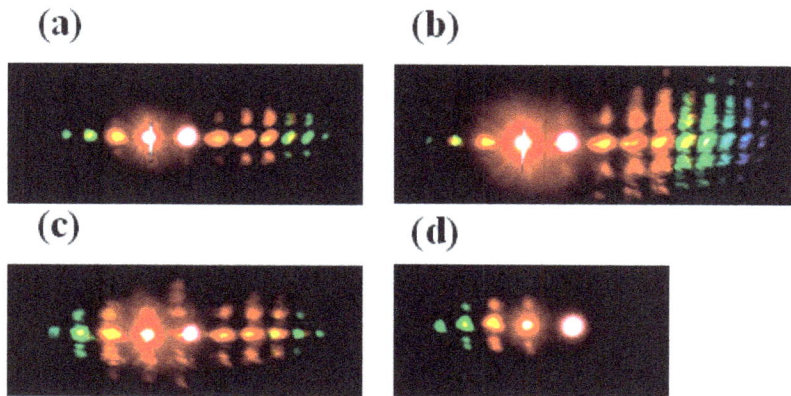

Figure 15. Photographs of the multicolored arrays generated with (**a**) pulse energy of beam 2 of 220 μJ, (**b**) pulse energy of beam 2 of 250 μJ; (**c**) time delay of two pump beams of 7 fs and pulse energy of beam 2 of 250 μJ; and (**d**) a short-pass filter cut-off wavelength at 820 nm inserted in the Beam 1 path [55].

We performed another experiment to study the characteristics of the 2-D multicolored arrays in detail. The schematic of this experiment is shown in Figure 16a. The two pump beams had a cross-angle of 1.8°, and a beam size of 300 μm in sapphire plate. Stable 2-D multicolored arrays were generated when Beam 1 and Beam 2 overlapped in time and space in the sapphire plate, as shown in Figure 16b. Spatially well-separated multicolored sidebands with >10 columns and rows were observed. The columns were approximately normal to the center row while the rows adjacent to the center row were not parallel to each other. The 2-D multicolored array sidebands are defined as $B_{m,n}$ for convenience, as shown in Figure 16c. $B_{0,0}$ and $B_{-1,0}$ stand for the two pump beams, Beam 1 and Beam 2, respectively.

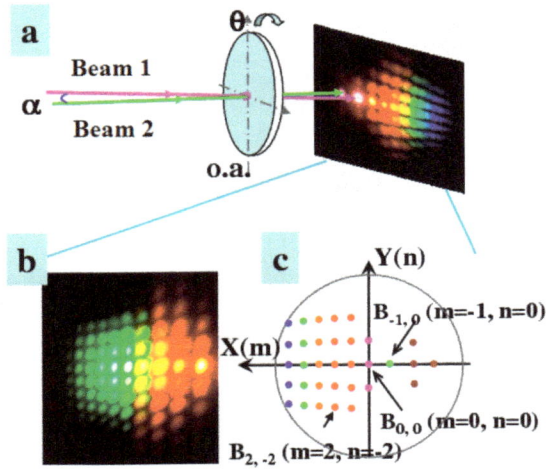

Figure 16. (a) Schematics of the generation of 2-D multicolored arrays; (b) A photograph of the 2-D multicolored arrays generated in sapphire plate; (c) 2-D multicolored arrays are defined as $B_{m,n}$, in which $B_{0,0}$ and $B_{-1,0}$ refer to Beam 1 and Beam 2 respectively [56]. o.a , optical axis.

The spectra of the sidebands on the center row $B_{m,0}$ and the second column $B_{2,n}$ were measured, as shown in Figure 17. The sidebands on the central row were generated through a CFWM process, which is almost the same as discussed in previous sections. The center wavelengths between neighboring spots on the same column were approximately the same, as shown in Figure 17b. A more accurate experiment using pump beams with narrower spectra will help to confirm these characteristics.

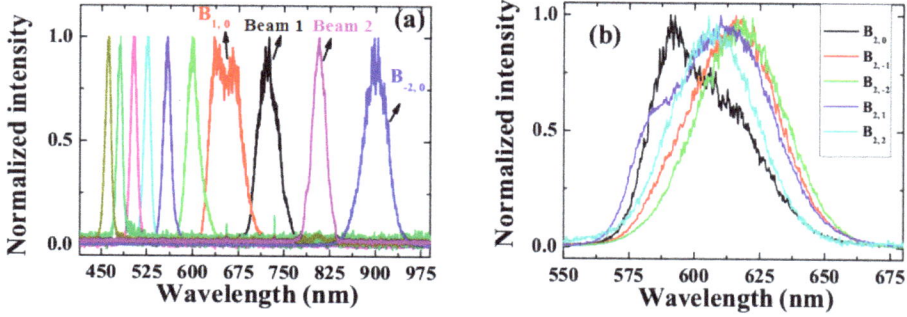

Figure 17. The spectra of sidebands on (**a**) the center row $B_{m,0}$; and (**b**) the second column $B_{2,n}$ [56].

The powers of Beam 1 and Beam 2 were set to 0.1 and 25 mW, respectively. The measured powers of some sidebands at this pump rate are shown in Figure 18a. The powers of sidebands on the rows of $B_{m,0}$, $B_{m,1}$, $B_{m,-1}$, $B_{m,2}$, $B_{m,-2}$ are shown as the star, red circle, black square, green triangle and blue triangle, respectively. We found that the power of the sidebands on $B_{m,1}$ and $B_{m,-1}$ were approximately the same as the value of m. The sidebands in $B_{m,2}$ and $B_{m,-2}$ also had this property, showing that the power distribution had mirror symmetry with the central line of $B_{m,0}$. The power dependence of three sidebands, $B_{1,0}$, $B_{2,0}$, and $B_{2,1}$, on the input power of Beam 2 are shown in the inset of Figure 18a. During the experiment, the power of Beam 1 was amplified from 0.1 to 0.17 mW, which means that the FWM process could also be used for parametric amplification. The power stability of several sidebands in different arrays, measured with a photodiode, is shown in Figure 18b. The stabilities are all in the range 0.5%–2% in RMS within 200 s.

In 2013, 2-dimensional multicolored arrays were observed with a low pump rate in a diamond plate, as shown in Figure 13. The experimental setup was the same as the Type-2 setup shown in Section 3.1. Increasing the pulse energy of Beam 2 to 0.856 µJ created 2-D multicolored arrays, as shown in Figure 13b. The threshold energy was much lower than that for a sapphire plate, *i.e.*, <2 µJ for a diamond plate and 25 µJ for a sapphire plate [56,62]. More sidebands, lines and brighter arrays were observed when the energy of Beam 2 was increased further, as shown in Figure 13c,d.

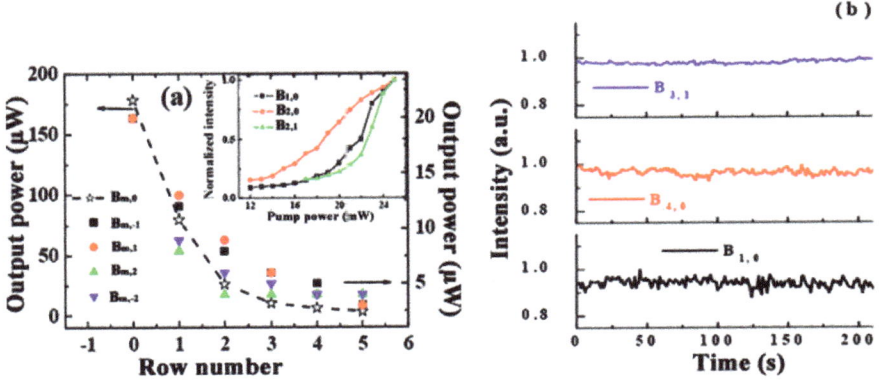

Figure 18. (a) The output power of sibebands on the $B_{m,0}$, $B_{m,1}$, $B_{m,-1}$, $B_{m,-2}$, $B_{m,2}$, with pump power of 0.1 mW and 25 mW for Beam 1 and Beam 2, respectively. The inset shows the power dependence of three different sidebands, $B_{1,0}$, $B_{2,0}$, and $B_{2,1}$, on power of Beam 2; (b) The power stabilities of three sidebands $B_{3,1}$, $B_{4,0}$ and $B_{1,0}$ [56].

The mechanism for the generation of 2D structure is not yet fully clear, although we have given some explanations based on the CFWM and beam breakup [62]. The detailed simulations of spontaneous breakup of elliptical laser beams were performed by Majus and his coworkers [76]. They attributed this breakup to multistep four-wave and parametric amplification of certain components occurring in the spatial spectrum of the self-focusing laser beam. Interestingly, beam breakup was observed even with a near circular (the ellipticity $e = 1.09$) input beam when the input power was ~20 times larger than P_{cr}, which is defined as follows [77]:

$$P_{cr} = 3.77\lambda^2 / (8\pi n n_2) \tag{4.1}$$

Here, λ is the laser wavelength in vacuum, n is the refractive index, and n_2 is the nonlinear refractive index. The pump beams have an ellipticity of ~1.2 in the experiment, due to asymmetric focusing with several concave mirrors. The summation of the peak powers of the two input beams is about ~60 P_{cr} ($P_{cr} = 0.4$ MW for diamond plate), and would be large enough for beam breakup [76]. The combination of this beam breakup and CFWM is the main cause of this 2-D multicolor structure.

Beam breakup can also occur when the pump beams are circular. Dergachev has investigated the interaction of two noncollinear femtosecond laser filaments in sapphire both numerically and experimentally [78]. The simulation was based on the nonlinear Schrödinger equation. The experiment was performed with a Ti:sapphire amplifier with a pulse width of 120 fs. The two incident beams, which are

24

all from the same laser source, have a cross-angle of 4.64°. Because of the asymmetric distributions of refractive index changes in the nonlinear material due to the Kerr effect of the pump beams, additional "hot points" or plasma channels arise in the plane oriented perpendicular to the pulse propagation plane with input pulse power above 10 P_{cr} (corresponding to a pulse energy of 3~4 µJ). Clear beam breakup was observed. The energy distribution of light spots at the output end can be adjusted by tuning the phase between the two input beams.

We are working on simulations based on the nonlinear Schrödinger equation including FWM and other nonlinear processes, such as optical Kerr effect and multiphoton absorption, which are required for full understanding of this phenomenon.

5. Conclusions and Prospects

In conclusion, we have investigated multicolor sideband generation based on CFWM both theoretically and experimentally. Analysis and computer simulation including the effect of phase-matching in FWM reasonably explain the reported experimental results.

The main characteristics of multicolored sidebands obtained in our experiment can be summarized as follows.

(1) Tunability in a wide spectral region.
   Fifteen spectral up-shifted pulses and two spectral down-shifted emissions were obtained simultaneously in a spectra domain that spanned more than 1.8 octaves. The wavelengths of the sidebands could be tuned from near-ultraviolet to near-infrared by adjusting the crossing angle between the two input beams or by replacing the nonlinear bulk medium.

(2) Ultrashort pulse width.
   The pulse width of the sidebands remained nearly unchanged for the Stokes and anti-Stokes pulses. Nearly transform-limited compressed pulses as short as 15 fs could be obtained when one of the two input beams was properly negatively chirped and the other was positively chirped.

(3) High output energy.
   The pulse energy of the sideband could be increased to 1 µJ, a power stability better than 1% RMS. We expect that an even higher output power could be generated by increasing the pump energy and expanding the spot sizes of the two pump beams on the optical medium to avoid saturation.

We have reported the generation of multicolor sidebands consisting of 2-D arrays, and provided some possible explanations. Beam breakup and CFWM are responsible for this interesting phenomenon. Careful investigation using both simulation and experiment are still needed for complete understanding of this new phenomenon.

Future studies into multicolored sidebands extending in the visible and near-IR spectral regions, which are generated with pump lasers with MHz repetition rates, would be useful in numerous applications such as nonlinear microscopy. A pulse energy of 1 μJ for each pump pulse has been confirmed to be enough energy for multicolor sideband generation in a diamond plate. This energy can be reduced further, if the pump beams are tightly focused on a medium with higher third-order nonlinearity. For example, CdSSe-nanoparticle-doped glass has 5.6 times larger third-order susceptibility than a diamond plate, even at wavelengths far from its resonant frequency [73].

Acknowledgments: This work was partly supported by the 21st Century COE program on "Coherent Optical Science", and partly supported by the grant from the Ministry of Education (MOE) in Taiwan under the ATU Program at National Chiao Tung University. A part of this work was performed as a joint research project with the Laser Engineering group at Osaka University, under contract subject B1-27.

Author Contributions: Jun Liu performed most of the experiments; Jinping He performed part of the experiments and wrote the paper; Takayoshi Kobayashi proposed some of the experiments, discussed with the coauthors, and also revised the paper.

Conflicts of Interest: The authors declare no conflict of interest.

References

1. Dantus, M.; Rosker, M.J.; Zewail, A.H. Real-time femtosecond probing of "transition states" in chemical reactions. *J. Chem. Phys.* **1987**, *87*, 2395.
2. Bowman, R.M.; Dantus, M.; Zewail, A.H. Femtosecond transition-state spectroscopy of iodine: From strongly bound to repulsive surface dynamics. *Chem. Phys. Lett.* **1989**, *161*, 297–302.
3. Douhal, A.; Kim, S.K.; Zewail, A.H. Femtosecond molecular dynamics of tautomerization in modelbase pairs. *Nature* **1995**, *378*, 260–263.
4. Hertel, V.; Raldoff, W. Ultrafast dynamics in isolated molecules and molecular clusters. *Rep. Prog. Phys.* **2006**, *69*, 1897.
5. Kobayashi, T.; Saito, T.; Ohtani, H. Real-time spectroscopy of transition states in bacteriorhodopsin during retinal isomerization. *Nature* **2001**, *414*, 531–534.
6. Kobayashi, T.; Kida, Y. Ultrafast spectroscopy with sub-10fs deep-ultraviolet pulses. *Phys. Chem. Chem. Phys.* **2012**, *14*, 6200–6210.
7. Middleton, C.T.; Harpe, K.L.; Su, C.; Law, Y.K.; Hernandez, C.E.C.; Kohler, B. DNA excited-state dynamics: From single bases to double Helix. *Annu. Rev. Phys. Chem.* **2009**, *60*, 217–239.
8. Denk, W.; Strickler, J.H.; Webb, W.W. Two-photon laser scanning fluorescence microscopy. *Science* **1990**, *248*, 73–76.

9. Helmchen, F.; Denk, W. Deep tissue two-photon microscopy. *Nat. Methods* **2005**, *2*, 932–940.

10. Horton, N.G.; Wang, K.; Kobat, D.; Clark, C.W.; Wise, F.W.; Schaffer, C.B.; Xu, C. *In vivo* three-photon microscopy of subcortical structures within an intact mouse brain. *Nat. Photon.* **2013**, *7*, 205–209.

11. Campagnola, P.J.; Loew, L.M. Second-harmonic imaging microscopy for visualizing biomolecular arrays in cells, tissues and organisms. *Nat. Biotechnol.* **2003**, *21*, 1356–1360.

12. Campagnola, P.J.; Clark, H.A.; Mohler, W.A. Second-harmonic imaging microscopy of living cells. *J. Biomed. Opt.* **2001**, *6*, 277–286.

13. Débarre, D.; Supatto, W.; Pena, A.M.; Fabre, A.; Tordjmann, T.; Combettes, L.; Schanne-Klein, M.C.; Beaurepair, E. Imaging lipid bodies in cells and tissues using third-harmonic generation microscopy. *Nat. Methods* **2006**, *3*, 47–53.

14. Barad, Y.; Eisenberg, H.; Horowitz, M.; Silberger, Y. Nonlinear scanning laser microscopy by third harmonic generation. *Appl. Phys. Lett.* **1997**, *70*, 922.

15. Gattass, R.R.; Mazur, E. Femtosecond laser micromachining in transparent materials. *Nat. Photon.* **2008**, *2*, 219–225.

16. Liu, X.; Du, D.; Mourou, G. Laser ablation and micromachining with ultrashort laser pulses. *IEEE J. Quant. Electron.* **1997**, *33*, 1706–1716.

17. Schaffer, C.B.; Brodeur, A.; Garcia, J.F.; Mazur, E. Micromachining bulk glass by use of femtosecond laser pulses with nanojoule enrgy. *Opt. Lett.* **2001**, *26*, 93–95.

18. Huang, M.; Zhao, F.L.; Cheng, Y.; Xu, N.S.; Xu, Z.Z. Origin of laser-induced near-subwavelength ripples: Interference between surface plasmons and incident laser. *ACS Nano* **2009**, *3*, 4062–4070.

19. Zgadzaj, R.; Gaul, E.; Matlis, N.H.; Shvets, G.; Downer, M.C. Femtosecond pump-probe study of preformed plasma channels. *J. Opt. Soc. Am. B* **2004**, *21*, 1559–1567.

20. Hochstrasser, R.M. Two-dimensional spectroscopy at infrared and optical frequencies. *Proc. Natl. Acad. Sci. USA* **2007**, *104*, 14190–14196.

21. Dunn, K.W.; Sandoval, R.M.; Kelly, K.J.; Dagher, P.C.; Tanner, G.A.; Atkinson, S.J.; Bacallao, R.L.; Molitoris, B.A. Functional studies of the kidney of living animals using multicolor two-photon microscopy. *Am. J. Physiol. Cell Physiol.* **2002**, *283*, 905–916.

22. Sahai, E.; Wyckoff, J.; Philippar, U.; Segall, J.E.; Gertler, F.; Condeelis, J. Simultaneous imaging of GFP, CFP and collagen in tumors *in vivo* using multiphoton microscopy. *BMC Biotechnol.* **2005**, *5*, 14.

23. Mahou, P.; Zimmerley, M.; Loulier, K.; Matho, K.S.; Labroille, G.; Morin, X.; Supatto, W.; Livet, J.; Débarre, D.; Beaurepaire, E. Multicolor two-photon tissue imaging by wavelength mixing. *Nat. Methods* **2012**, *9*, 815–818.

24. Frank, P.A.; Hill, A.E.; Peters, C.W.; Weinreich, G. Generation of optical harmonics. *Phys. Rev. Lett.* **1961**, *7*, 118–120.

25. Seifert, F.; Ringling, J.; Noack, F.; Petrov, V.; Kittelmann, O. Generation of tunable femtosecond pulses to as low as 172.7 nm by sum-frequency mixing in lithium triborate. *Opt. Lett.* **1994**, *19*, 1538–1540.

26. Liu, J.; Kida, Y.; Teramoto, T.; Kobayashi, T. Generation of stable sub-10fs pulses at 400 nm in a hollow fiber for UV pump-probe experiment. *Opt. Express* **2010**, *18*, 4664–4672.

27. Baum, P.; Lochbrunner, S.; Riedle, E. Tunable sub-10-fs ultraviolet pulses generated by achromatic frequency doubling *Opt. Lett.* **2004**, *68*, 2793–2795.

28. Zhao, B.; Jiang, Y.; Sueda, K.; Miyanaga, N.; Kobayashi, T. Sub-15fs ultraviolet pulses generated by achromatic phase-matching sum-frequency mixing. *Opt. Express* **2009**, *17*, 17711–17714.

29. Aközbek, N.; Iwasaki, A.; Becker, A.; Chin, S.L.; Bowden, C.M. Third-harmonic generation and self-channeling in air using high-power femtosecond laser pulses. *Phys. Rev. Lett.* **2002**, *89*, 143901.

30. Tzankov, P.; Steinkellner, O.; Zheng, J.; Mero, M.; Freyer, W.; Husakou, A.; Babushkin, I.; Herrmann, J.; Noack, F. High-power fifth-harmonic generation of femtosecond pulses in vacuum ultraviolet using a Ti: Sapphire laser. *Opt. Express* **2007**, *15*, 6389–6395.

31. Macklin, J.J.; Kmetec, J.D.; Gordon, C.L. High-order harmonic generation using intense femtosecond pulses. *Phys. Rev. Lett.* **1993**, *70*, 766.

32. Giordmaine, J.A.; Miller, R.C. Tunable coherent parametric oscillation in LiNbO$_3$ at optical frequencies. *Phys. Rev. Lett.* **1965**, *14*, 973–976.

33. Edelstein, D.C.; Wachman, E.S.; Tang, C.L. Broadly tunable high repetition rate femtosecond parametric oscillator. *Appl. Phys. Lett.* **1989**, *54*, 1728.

34. Gale, G.M.; Cavallari, M.; Driscoll, T.J.; Hasche, F. Sub-20-fs tunable pulses in the visible from an 82-MHz optical parametric oscillator. *Opt. Lett.* **1995**, *20*, 1562–1564.

35. Burr, K.C.; Tang, C.L.; Arbore, M.A.; Fejer, M.M. Broadly tunable mid-inrared femtosecond optical parametric oscillator using all-solid-state-pumped periodically poled lithium niobate. *Opt. Lett.* **1997**, *22*, 1458–1460.

36. Wilhelm, T.; Piel, J.; Riedle, E. Sub-20fs tunable across the visible from blue-pumped single-pass nonlinear parametric converter. *Opt. Lett.* **1997**, *22*, 1494–1496.

37. Cerullo, G.; Nisoli, M.; Stagira, S.; de Silvestri, S. Sub-8-fs pulses from an ultrabroadband optical parametric amplifier in the visible. *Opt. Lett.* **1998**, *23*, 1283–1285.

38. Shirakawa, A.; Sakane, I.; Kobayashi, T. Pulse-front-matched optical parametric amplification for sub-10-fs pulse generation tunable in the visible and near infrared. *Opt. Lett.* **1998**, *23*, 1292–1294.

39. Okamura, K.; Kobayashi, T. Octave-spanning carrier-envelope phase stabilized visible pulse with sub-3-fs pulse duration. *Opt. Lett.* **2011**, *36*, 226–228.

40. Shirakawa, A.; Sakane, I.; Takasaka, M.; Kobayashi, T. Sub-5-fs visible pulse generation by pulse-front-matched noncollinear optical parametric amplification. *Appl. Phys. Lett.* **1999**, *74*, 2268–2270.

41. Corkum, P.B.; Rolland, C.; Srinivasan-Rao, T. Supercontinuum generation in gases. *Phys. Rev. Lett.* **1986**, *57*, 2268.

42. Kasparian, J.; Sauerbrey, R.; Mondelain, D.; Niedermeier, S.; Yu, J.; Wolf, J.P.; André, Y.B.; Franco, M.; Prade, B.; Tzortzakis, S.; *et al.* Infrared extension of super continuum generated by femtosecond terawatt laser pulses propagating in the atmosphere. *Opt. Lett.* **2000**, *25*, 1397–1399.

28

43. Kandidov, V.P.; Kosareva, O.G.; Golubtsov, I.S.; Liu, W.; Becker, A.; Akozbek, N.; Bowden, C.M.; Chin, S.L. Self-transformation of a powerful femtosecond laser pulse into a white-light laser pulse in bulk optical media(or supercontinuum generation). *Appl. Phys. B* **2003**, *77*, 149–165.

44. Wadsworth, W.J.; Blanch, A.O.; Knight, J.C.; Birks, T.A.; Martin Man, T.P.; Russell, P.S.J. Supercontinuum generation in photonic crystal fibers and optical fiber tapers: A novel light source. *JOSA B* **2002**, *19*, 2148–2155.

45. Xia, C.; Kumar, M.; Kulkarni, O.P.; Islam, M.N.; Terry, F.L.; Freeman, J.M.J.; Poulain, M.; Mazé, G. Mid-infrared supercontinuum generation to 4.5 μm in ZBLAN fluoride fibers by nanosecond diode pumping. *Opt. Lett.* **2006**, *31*, 2553–2555.

46. Dunsby, C.; Lanigan, P.M.P.; McGinty, J.; Elson, D.S.; Isidro, J.R.; Munro, I.; Galletly, N.; McCann, F.; Treanor, B.; Önfelt, B.; *et al.* An electronically tunable ultrafast laser source applied to fluorescence imaging and fluorescence lifetime imaging microscopy. *J. Phys. D* **2004**, *37*, 3296–3303.

47. Gu, X.; Xu, L.; Kimmel, M.; Zeek, E.; O'Shea, P.; Shreenath, A.P.; Trebino, R. Frequency-resolved optical gating and single-shot spectral measurements reveal fine structure in microstructure-fiber continuum. *Opt. Lett.* **2000**, *27*, 1174–1176.

48. Dudley, J.M.; Genty, G.; Coen, S. Supercontinuum generation in photonic crystal fiber. *Rev. Mod. Phys.* **2006**, *78*, 1135–1184.

49. Fuji, T.; Horio, T.; Suzuki, T. Generation of 12 fs deep-ultraviolet pulses by four-wave mixing through filamentation in neon gas. *Opt. Lett.* **2007**, *32*, 2481–2483.

50. Okamoto, H.; Tatsumi, M. Generation of ultrashort light pulses in the mid-infrared (3000–800 cm^{-1}) by four-wave mixing. *Opt. Commun.* **1995**, *121*, 63–68.

51. Fuji, T.; Suzuki, T. Generation of sub-two-cycle mid-infrared pulses by four-wave mixing through filamentation in air. *Opt. Lett.* **2007**, *32*, 3330–3332.

52. Kida, Y.; Liu, J.; Teramoto, T.; Kobayashi, T. Sub-10fs deep-ultraviolet pulses generated by chirped-pulse four-wave mixing. *Opt. Lett.* **2010**, *35*, 1807–1809.

53. He, J.; Kobayashi, T. Generation of sub-20fs deep-ultraviolet pulses by using chirped-pulse four-wave mixing in CaF$_2$ plate. *Opt. Lett.* **2013**, *38*, 2938–2940.

54. Crespo, H.; Mendonça, J.T.; Dos Santos, A. Cascaded highly nondegenerate four-wave-mixing phenomenon in transparent isotropic condensed media. *Opt. Lett.* **2000**, *25*, 829–831.

55. Liu, J.; Kobayashi, T. Cascaded four-wave mixing and multicolored arrays generation in a sapphire plate by using two crossing beams of femtosecond laser. *Opt. Express* **2008**, *16*, 22119–22125.

56. Liu, J.; Kobayashi, T.; Wang, Z.G. Generation of broadband two-dimensional multicolored arrays in a sapphire plate. *Opt. Express* **2009**, *17*, 9226–9234.

57. Liu, J.; Zhang, J.; Kobayashi, T. Broadband coherent anti-Stokes Raman scattering light generation in BBO crystal by using two crossing femtosecond laser pulses. *Opt. Lett.* **2008**, *33*, 1494–1496.

58. Liu, W.; Zhu, L.; Fang, C. Observation of sum-frequency-generation-induced cascaded four-wave mixing using two crossing femtosecond laser pulse in a 0.1 mm beta-barium-borate crystal. *Opt. Lett.* **2012**, *37*, 3783–3785.

59. Liu, J.; Kobayashi, T. Wavelength-tunable multicolored femtosecond laser pulse generation in fused silica glass. *Opt. Lett.* **2009**, *34*, 1066–1068.

60. Liu, J.; Kobayashi, T. Cascaded four-wave mixing in transparent bulk media. *Opt. Comm.* **2010**, *283*, 1114–1123.

61. Zhi, M.; Wang, X.; Sokolov, A.V. Broadband coherent light generation in diamond driven by femtosecond pulses. *Opt. Express* **2008**, *16*, 12139–12147.

62. He, J.; Du, J.; Kobayashi, T. Low-threshold and compact multicolored femtosecond laser generated by using cascaded four-wave mixing in a diamond plate. *Opt. Comm.* **2013**, *290*, 132–135.

63. Liu, J.; Kobayashi, T. Generation of sub-20-fs multicolor laser pulses using cascaded four-wave mixing with chirped incident pulses. *Opt. Lett.* **2009**, *34*, 2402–2404.

64. Weigand, R.; Mendonca, J.T.; Crespo, H. Cascaded nondegenerate four-wave mixing technique for high-power single-cycle pulse synthesis in the visible and ultraviolet ranges. *Phys. Rev. A* **2009**, *79*, 063838.

65. Silva, J.L.; Weigand, R.; Crespo, H. Octave-spanning spectra and pulse synthesis by non-degenerate cascaded four-wave mixing. *Opt. Lett.* **2009**, *34*, 2489–2491.

66. Liu, J.; Kobayashi, T. Generation of µJ-level multicolored femtosecond laser pulses using cascaded four-wave mixing. *Opt. Express* **2009**, *17*, 4984–4990.

67. Nietzke, R.; Fenz, P.; Elsässer, W.; Göbel, E.O. Cascaded fourwave mixing in semiconductor laser. *Appl. Phys. Lett.* **1987**, *51*, 1298–1300.

68. Eckbreth, A.C.; Anderson, T.J.; Dobbs, G.M. Multi-color CARS for Hydrogen-fueled scramjet applications. *Appl. Phys. B* **1988**, *45*, 215–223.

69. Sokolov, A.V.; Walker, D.R.; Yavuz, D.D.; Yin, G.Y.; Harris, S.E. Raman generation by phased and antiphased molecular states. *Phys. Rev. Lett.* **2000**, *85*, 562–565.

70. Zhang, H.; Liu, H.; Si, J.; Yi, W.; Chen, F.; Hou, X. Low threshold power density for the generation of frequency up-converted pulses in bismuth glass by two crossing chirped femtosecond pulses. *Opt. Express* **2011**, *19*, 12039–12044.

71. Zeng, H.; Wu, J.; Xu, H.; Wu, K. Generation and weak beam control of two-dimensional multicolored arrays in a quadratic nonlinear medium. *Phys. Rev. Lett.* **2006**, *96*, 083902.

72. Liu, W.; Zhu, L.; Fang, C. *In-situ* weak-beam and polarization control of multidimensional laser sidebands for ultrafast optical switching. *Appl. Phys. Lett.* **2014**, *104*, 111114.

73. Boyd, R.W. *Nonlinear Optics*, 3rd ed.; Elsevier: Singapore, 2010.

74. Liu, J.; Kida, Y.; Teramoto, T.; Kobayashi, T. Simultaneous compression and amplification of a laser pulse in a glass plate. *Opt. Express* **2010**, *18*, 2495–2502.

75. Liu, J.; Kobayashi, T. Generation and amplification of tunable multicolored femtosecond laser pulses by using cascaded four-wave mixing in transparent bulk media. *Sensors* **2010**, *10*, 4296–4341.

76. Majus, D.; Jukna, V.; Valiulis, G.; Dubietis, A. Generation of periodic filament arrays by self-focusing of highly elliptical ultrashort pulsed laser beams. *Phys. Rev. A* **2009**, *79*, 033843.

77. Dubietis, A.; Tamosauskas, G.; Fibich, G.; Ilan, B. Multiple filamentation induced by input beam elliticity. *Opt. Lett.* **2004**, *29*, 1451–1453.

78. Dergachev, A.A.; Kadan, V.N.; Shlenov, S.A. Interaction of noncolinear femtosecond laser filaments in sapphire. *Quant. Electron.* **2012**, *42*, 125–129.

Fundamentals of Highly Non-Degenerate Cascaded Four-Wave Mixing

Rosa Weigand and Helder M. Crespo

Abstract: By crossing two intense ultrashort laser pulses with different colors in a transparent medium, like a simple piece of glass, a fan of multicolored broadband light pulses can be simultaneously generated. These newly generated pulses are emitted in several well-defined directions and can cover a broad spectral range, from the infrared to the ultraviolet and beyond. This beautiful phenomenon, first observed and described 15 years ago, is due to highly-nondegenerate cascaded four-wave mixing (cascaded FWM, or CFWM). Here, we present a review of our work on the generation and measurement of multicolored light pulses based on third-order nonlinearities in transparent solids, from the discovery and first demonstration of highly-nondegenerate CFWM, to the coherent synthesis of single-cycle pulses by superposition of the multicolored light pulses produced by CFWM. We will also present the development and main results of a dedicated 2.5-D nonlinear propagation model, *i.e.*, with propagation occurring along a two-dimensional plane while assuming cylindrically symmetric pump beam profiles, capable of adequately describing noncollinear FWM and CFWM processes. A new method for the generation of femtosecond pulses in the deep-ultraviolet (DUV) based on FWM and CFWM will also be described. These experimental and theoretical results show that highly-nondegenerate third-order nonlinear optical processes are formally well understood and provide broader bandwidths than other nonlinear optical processes for the generation of ultrashort light pulses with wavelengths extending from the near-infrared to the deep-ultraviolet, which have many applications in science and technology.

Reprinted from *Appl. Sci.* Cite as: Weigand, R.; Crespo, H.M. Fundamentals of Highly Non-Degenerate Cascaded Four-Wave Mixing. *Appl. Sci.* **2015**, *5*, 485–515.

1. Introduction

Four-wave mixing (FWM) processes result from the interplay between four electromagnetic waves coupled through the optical Kerr nonlinearity of a medium, given by the third-order susceptibility $\chi^{(3)}$ [1]. Since all media, either isotropic or anisotropic (with any kind of crystal symmetry), have a non-zero third order susceptibility, FWM processes have a universal character. The four interacting waves can all have the same frequency, in which case the process is degenerate and the corresponding susceptibility is given by $\chi^{(3)}(\omega;\omega,\omega,-\omega)$. In this case, three fields

32

with frequency ω are mixed in the medium to produce another field also with frequency ω. The energy conservation law that determines the frequency of the newly generated field is given in this case by: $\omega = \omega + \omega - \omega \equiv 2\omega - \omega$. In practice, this process can be observed by crossing two intense pulses in a nonlinear medium at a small angle. For incident wavevectors k_0 and k_1, momentum conservation dictates that the newly generated beams will be emitted in the directions given by $k_2 = 2k_1 - k_0$ and $k'_2 = 2k_0 - k_1$, as shown in Figure 1. This process can be seen as the diffraction of the incident beams by the nonlinear index grating (laser induced grating) produced by the same beams [2]. In bulk dispersive media this process is intrinsically phase mismatched (see, e.g., [3]), with a mismatch given approximately by $\delta k = 2k(1 - \cos\theta) \simeq k\theta^2$, where $k = |k_0| = |k_1|$. This means that the efficiency of the process drops for increasing interaction (and hence emission) angles.

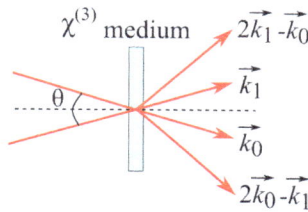

Figure 1. Basic geometry of nonlinear self-diffraction.

The waves can also have different frequencies, so a new field with frequency ω_2 can be generated departing from two fields with frequencies ω_0 and ω_1 ($\omega_1 > \omega_0$). In this case, the nonlinear polarization terms which describe diffraction from a moving laser-induced grating give rise to frequency upshifted and downshifted pulses with frequencies $\omega_2 = 2\omega_1 - \omega_0$ and $\omega'_2 = 2\omega_0 - \omega_1$, respectively, so the output consists of four beams with different colors emitted in different directions. Penzkofer and Lehmeier [4] analized theoretically the noncollinear phase-matched amplification of ultrashort light pulses via four-wave mixing in isotropic media, having derived explicit expressions for the phase-matching angles and gain.

If the fields are intense enough, the third order susceptibility $\chi^{(3)}$ can continue mixing the fields, as long as the energy and momentum conservation laws are fulfilled for the new processes. This way, new fields are generated via cascaded four-wave mixing (CFWM). For noncollinear interactions, this gives rise to additional beams that can be seen as higher orders of diffraction of the moving grating (note that the moving grating corresponds to the spatially and temporally varying nonlinear refractive index change inside the medium, since the medium itself is stationary in the laboratory frame). For the degenerate case, the larger the interaction angle, the smaller the observed number of diffracted orders (due to increasing phase-mismatch), but for the nondegenerate case we will see that the phase-mismatch in cascaded processes can be reduced by proper choice of the interaction angle between the two initial beams.

Collinear fifth-harmonic generation in a crystal as a result of cascaded processes of a cubic nonlinearity was reported in an early work [5]. Extensive work has been done in this field by mixing near-degenerate nanosecond laser pulses in optical fibers and obtaining several pairs of sidebands [6–8]. Cascaded processes can also occur with second order processes and have been observed in BBO crystals [9] or in an hybrid form with third order processes in picosecond optical parametric amplifiers [10].

Further progress was made when highly nondegenerate CFWM was demonstrated by mixing femtosecond visible laser pulses with different colors. The first demonstration was done in bulk isotropic media (thin glass slide) [11], and resulted in the generation of multiple broadband light pulses extending from the infrared to the ultraviolet starting from two femtosecond laser pulses in the visible range. Nonresonant nondegenerate CFWM in the femtosecond regime was subsequently observed and demonstrated in other nonlinear media and spectral regions, from semiconductors pumped with mid-IR pulses [12] to gases and plasmas pumped in the NIR to uv. In particular, Misoguti et al. [13] successfully extended CFWM to the vacuum-uv range (down to 160 nm) using a gas-filled hollow waveguide pumped with ω and 2ω pulses from a Ti:Sapphire laser. Raman-assisted noncollinear CFWM has also been obtained in Raman-active media, such as diamond, using dual color laser pulses and chirped broadband pulses (see, e.g., [14] and references therein).

This paper gives a review of the work done by the authors in this field and includes the experimental procedures and a complete theoretical model explaining not only the spectral characteristics of the generated beams, but also their intensity, spectral phases, emission angles and energies. Important examples of the application of the generated pulses namely for the synthesis of single cycle pulses, will be discussed, as well as the possibility of using CFWM for generating broadband light pulses in the deep ultraviolet spectral region.

2. Basic Experiment and Interpretation

The first experiment on highly nondegenerate noncollinear CFWM was done by H. Crespo et al. at LOA (Laboratoire d'Optique Appliquée in Palaiseau, France) [11] and the experimental setup can be seen in Figure 2a. The laser source is a dye laser-amplifier system delivering an orange beam (frequency ω_0, $\lambda_0 = 618$ nm, 80 fs) and a green beam (frequency ω_1, $\lambda_1 = 561$ nm, 40 fs). Both beams are horizontally polarized and with near-Gaussian spatial profiles (approximately 5 mm diameter). The green beam was sent through a delay-line and both beams were coupled through a 50/50 beam splitter to impinge on a large section plane-convex lens (f = 30 cm) at an adjustable small angle. The lens focuses both beams in a thin slide of BK7 glass with 150 μm thickness at an external angle θ = 2.9°. A thin glass slide was chosen to minimize self-phase modulation processes and it was also placed 1 cm before the focus of the lens to avoid laser-induced damage. At the plane of the slide both beams had 20 μJ energy, corresponding to intensities of 1.0 TW cm^{-2} for the orange and 2.1 TW cm^{-2} for the green beam. When both beams were temporally synchronized at the plane of the slide, a large set of multicolored beams (Figure 2b) was produced at both sides of the incident beams, consisting of two frequency-downshifted beams (we will call D_n the downconverted beam of order n) in the red and near infrared, as well as 11 frequency-upshifted beams (we will call U_n the upconverted beams of order n) extending into the blue and ultraviolet regions of the spectrum. The ultraviolet beams are seen as blue in the picture because the screen used was fluorescent, whereas the second downconverted beam could be observed with the help of a NIR viewer. The spectrum of the generated beams was registered by collecting the generated fan of beams with a large aperture aluminum-coated parabolic mirror M4 to minimize chromatic aberration and UV absorption and by sending the collimated beams to a spectrograph equipped with a CCD camera where the resulting spectrum is given in Figure 3. Two downconverted bands (D_1 and D_2) can be seen to the left of the pump beams, while only 5 upconverted bands (U_1 to U_5) are seen to the right, due to sensitivity limitations of the CCD in the blue and UV spectral ranges. All generated orders are broadband, with bandwidths of the order of those of the pump and signal beams (16 nm approximately) which are larger than the original laser bandwidths (5–6 mm) due to some self- and cross-phase modulation taking place within the slide. Energy measurements were done with a calibrated photodetector and the whole set of newly generated beams carried about 5%–10% of the total energy of the orange and green pulses.

The spectrum of Figure 3 can be explained, to a very good approximation, by assuming a cascaded FWM process in the central frequency approximation. The two pump pulses drive the $\chi^{(3)}$ medium at the modulation frequency $\omega_m = \omega_1 - \omega_0$, giving rise to multiple pulses with frequencies $\omega_n = \omega_0 + n\omega_m$, where n is the integer beam order and $n > 1$ ($n < 0$) denotes frequency upconverted (downconverted) pulses. This can also be written as $\omega_n = \omega_{n-1} + \omega_m$ or $\omega_n = n\omega_1 - (n - 1)\omega_0$. Note that even though the total number of photons involved in a given process is $2n$, this does not correspond to the order of the corresponding nonlinearities ($2n - 1$), but to that of an effective nonlinearity obtained via cascaded $\chi^{(3)}$, i.e., third-order, processes. Table 1 shows, in an equivalent way, how the frequencies of the different orders are generated from the previous order plus or minus the modulation frequency.

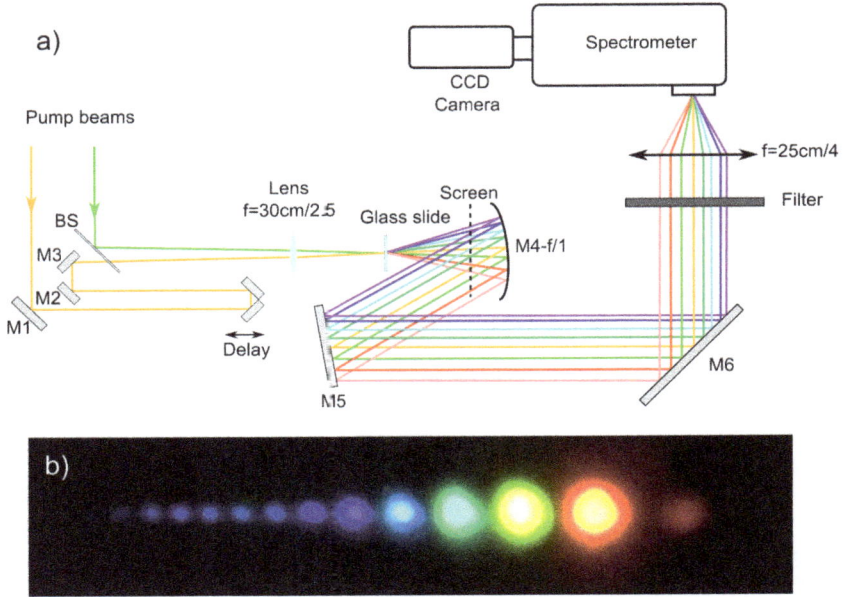

Figure 2. (a) Experimental arrangement: BS, beam splitter; M1–M3, silver mirrors; M4, parabolic mirror; M5, M6, aluminum mirrors. Spacing between colors not at scale; (b) Picture of output beams as taken on a white card. Adapted from [11].

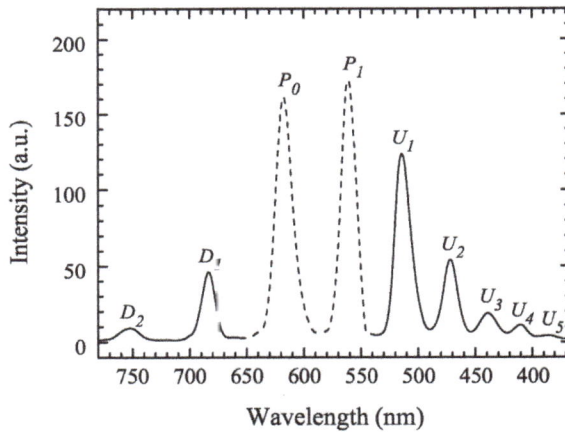

Figure 3. Spectra of the beams in the cascade: dashed curve, original spectra of the orange (P_0) and green (P_1) pump beams after crossing the glass slide but without temporal overlap (scaled for comparison); solid curve, spectra of the simultaneously generated beams. Adapted from [11].

36

Table 1. Approximate frequencies generated by the Cascaded Four-Wave Mixing process.

Order n	Frequency $\omega_n = \omega_0 - n\omega_m$	Equivalence
$-n$	$\omega_{-n} = \omega_0 - n\omega_m = (n+1)\omega_0 - n\omega_1$	$\omega_{-n} = \omega_{-n+1} - \omega_m$
.	.	.
.	.	.
.	.	.
-2	$\omega_{-2} = \omega_0 - 2\omega_m = 3\omega_0 - 2\omega_1$	$\omega_{-2} = \omega_{-1} - \omega_m$
-1	$\omega_{-1} = \omega_0 - \omega_m = 2\omega_0 - \omega_1$	$\omega_{-1} = \omega_0 - \omega_m$
0	$\omega_0 = \omega_0$	ω_0
1	$\omega_1 = \omega_0 + \omega_m = \omega_1$	ω_1
2	$\omega_2 = \omega_0 + 2\omega_m = 2\omega_1 - \omega_0$	$\omega_2 = \omega_1 + \omega_m$
3	$\omega_3 = \omega_0 + 3\omega_m = 3\omega_1 - 2\omega_0$	$\omega_3 = \omega_2 + \omega_m$
.	.	.
.	.	.
n	$\omega_n = \omega_0 + n\omega_m = n\omega_1 - (n-1)\omega_0$	$\omega_n = \omega_{n-1} + \omega_m$

An approximate description of the cascaded FWM process is completed by considering the phase-matching condition for the wave vectors of the upconverted and downconverted beams. Let us call k_0 and k_1 the wave vectors of the lower frequency (orange) and higher frequency (green) pump, whereas k_{-n} and k_n denote the wave vector of a downconverted beam or an upconverted beam of order n respectively. With $k_m = k_1 - k_0$ the modulation (or grating) wave vector, the newly generated wave vectors are approximately given by $k_n = k_0 + nk_m$, or equivalently by $k_n = k_{n-1} + k_m$ or $k_n = nk_1 - (n-1)k_0$. We will see that approximate phase matching can be achieved for a large number of frequency upconversion processes in a medium with normal dispersion, whereas frequency downconversion processes are intrinsically phase-mismatched, which explains the observed asymmetry between the two processes. For the same reason, the contribution of backwards processes such as $k_n = k_{n+1} - k_m$ is small and therefore can be neglected.

Table 2 shows how the wave vectors of the different orders are generated from the previous one plus or minus the modulation wave vector.

Figure 4 shows the diagrams for the vector additions in the cascaded FWM processes for both the upconverted and downconverted beams, along with their exiting angles β_n.

Table 2. Approximate wave vectors generated by the Cascaded Four-Wave Mixing process.

Order n	Wave Vector $k_n = k_0 + n k_m$	Equivalence
$-n$	$k_{-n} = k_0 - n k_m = (n+1)k_0 - n k_1$	$k_{-n} = k_{-n+1} - k_m$
.	.	.
.		.
.	.	
.		.
-2	$k_{-2} = k_0 - 2k_m = 3k_0 - 2k_1$	$k_{-2} = k_{-1} - k_m$
-1	$k_{-1} = k_0 - k_m = 2k_0 - k_1$	$k_{-1} = k_0 - k_m$
0	k_0	k_0
1	$k_1 = k_0 + k_m = k_1$	k_1
2	$k_2 = k_0 + 2k_m = 2k_1 - k_0$	$k_2 = k_1 + k_m$
3	$k_3 = k_0 + 3k_m = 3k_1 - 2k_0$	$k_3 = k_2 + k_m$
.	.	.
.		.
.	.	
n	$k_n = n k_1 - (n-1)k_0$	$k_n = k_{n-1} + k_m$

Moreover it is possible to calculate the interaction angle θ_n that the main beams should have inside the medium for phase matching to occur for a particular process of order n, under the assumed approximation. Let us choose a coordinate frame in which $k_1 = k_1 \times (1,0)$. In this frame k_0 would be given by $k_0 = k_0 (\cos\theta_n, \sin\theta_n)$ and let us have in mind that any wave vector is related with the refractive index of the medium through $k_i = n_r(\omega_i)\,\omega_i/c$. Thus for the upconverted beams the equation $k_n = n k_1 - (n-1)k_0$ is written in this coordinate system as

$$k_n = (n k_1 - (n-1)k_0 \cos\theta_n, (n-1)k_0 \sin\theta_n) \tag{1}$$

Its square modulus is given by

$$k_n^2 = n^2 k_1^2 - 2n(n-1)k_0 k_1 \cos\theta_n + (n-1)^2 k_0^2 \cos^2\theta_n + (n-1)^2 k_0^2 \sin^2\theta_n \tag{2}$$

and solving for $\cos(\theta_n)$ we obtain

$$\cos\theta_n = \frac{n^2 k_1^2 + (n-1)^2 k_0^2 - k_n^2}{2n(n-1)k_0 k_1} \tag{3}$$

which is valid for the upshifts ($n \geq 2$) and downshifts ($n \leq -1$). With reference to Figure 4a,b, the emission angles of the newly generated orders can be obtained by calculating the corresponding wave vectors (e.g., $k_2 = 2k_1 - k_0$ for the first frequency-upshifted beam), with the ideal phase-matching angle given by

Equation (3), and determining their angle with respect to the horizontal axis, which results in

$$\cos \beta_n = \frac{n^2 k_1{}^2 - (n-1)^2 k_0{}^2 - k_n{}^2}{2n(n-1)k_0 k_1} \tag{4}$$

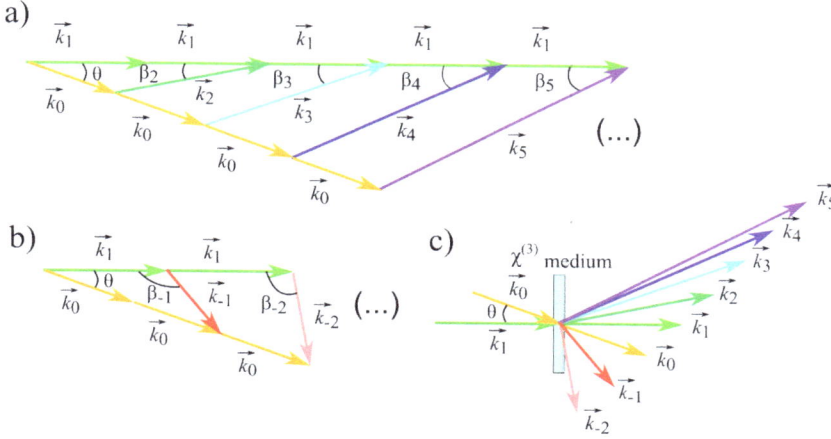

Figure 4. Wave vectors of the upconverted and downconverted beams generated by cascaded four-wave mixing (CFWM). (**a**) upconversion processes; (**b**) downconversion processes; (**c**) schematic of the wave vector distribution obtained by CFWM in real space.

Hence it is possible to calculate the interaction angle between the beams inside the medium for the different orders to appear. Thus one can chose for example the phase matching angle which optimizes the first upconverted order. This order will be generated in optimum energy conditions and the remaining orders will be generated less efficiently, since perfect phase-matching is not set for them. Also, it is not possible to exactly phase-match frequency down conversion processes in media with normal dispersion, as Equation (4) does not have a real solution, which results in the observed asymmetry between frequency upconverted and downconverted beams. The fact that the interaction angle affects the generation of the cascade of colors was experimentally observed [11] and the fact that angle phase-matching was achieved thanks to the material dispersion was settled, as detailed below. Figure 5 shows the measured and calculated wave vectors generated in the first nondegenerate CFWM experiment, where the two-color pump beams cross in the medium at an external angle of 2.9°. The arrows denote the pump beam wave vectors; the ten consecutive upconverted beams appear above the horizontal axis, and the two downconverted beams are below this axis. The open squares falling outside the dotted box are the five upconverted beams for which only the wave vector direction was measured directly, and hence the amplitude was extrapolated from the phase-matching conditions. The filled circles are calculated with the phase-matching conditions, where k_0 and k_1 were obtained directly from spectral and angular measurements of the pump pulses. The open circles are the calculations for perfectly phase-matched beams and are shown for reference only, since a cascade of such ideal processes (which would required different angles for each process) cannot be obtained experimentally.

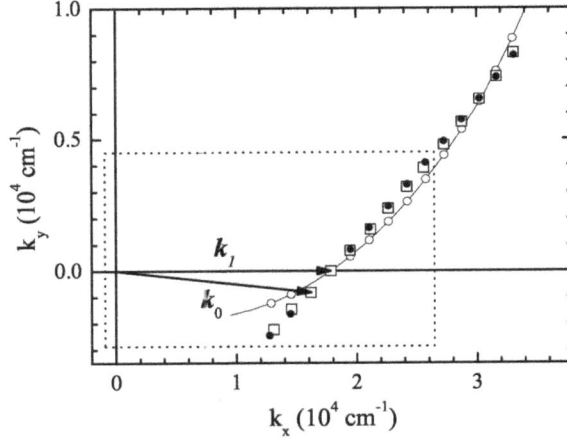

Figure 5. Representation of the wavevectors of all the beams in the pattern: open squares, measured points; filled circles, calculation with the approximate cascaded processes; open circles. calculated exactly phase-matched beams, with the interaction angle given by Equation (3). The solid curve is a guide. Adapted from [11].

Figure 6 shows the phase mismatch Δk_n for the optimized interaction angle of 2.9° as a function of beam order n, where $n > 1$ ($n < 0$) correspond to upconverted (downconverted) beams. The dashed curve is a calculation assuming collinear geometry (0°). As the interaction angle approaches 2.9°, the wave vector mismatch becomes clearly asymmetric, with the appearance of a region where the material dispersion is partially compensated for by the interaction angle. Since the self-phase modulated pump pulses have very large bandwidths, the geometry that results in phase-matching of the nth-order beam can give rise to multiple emission for all orders below n, provided that the maximum phase-mismatch for each order is less than approximately three times the bandwidth of the ultrafast pump pulses ($\sim$1350 cm^{-1}). We see that eleven upconverted beams and two downconverted beams all fall inside this region, and so the phase-matching conditions can be met in a quite relaxed way for several orders of the cascaded processes, within the frequency range allowed by the bandwidths, and in perfect agreement which the observed asymmetry between frequency upconversion and downconversion processes. Notice that, according to this criterion, a first frequency-upshifted beam should be generated even for a collinear interaction geometry (dashed curve of Figure 6). This is indeed the case, which has been confirmed by measuring the spectra of the resulting collinear and overlapping beams (with proper attenuation of the pump and signal pulses). Experimentally, the interaction angle is adjusted for maximum overall intensity and number of orders in the projected multicolored pattern. We expect the efficiency of a particular process to be determined by the magnitude of the nonlinear susceptibilities, the strength of the input fields, the phase-mismatch of any intermediate process, and ultimately absorption for any of the fields. Moreover, we see that the optimum yield does not necessarily occur under conditions that minimize the phase-mismatch for the final step in the process. Cascading will take place in the presence of some residual phase-mismatch in either the intermediate steps or the final step. The low degree of residual phase-mismatch also allows cascaded third-order processes to dominate over any high-order direct processes since cascaded processes will be favored if the phase-mismatch of the intermediate steps is sufficiently low. Storage of energy in the intermediate field then determines the conversion efficiency, similar to the case of an intermediate resonance. Although direct higher-order processes cannot be completely ruled out, they do not appear to take part in the observed phenomenon, as the detailed numerical simulations presented in the next section also demonstrate.

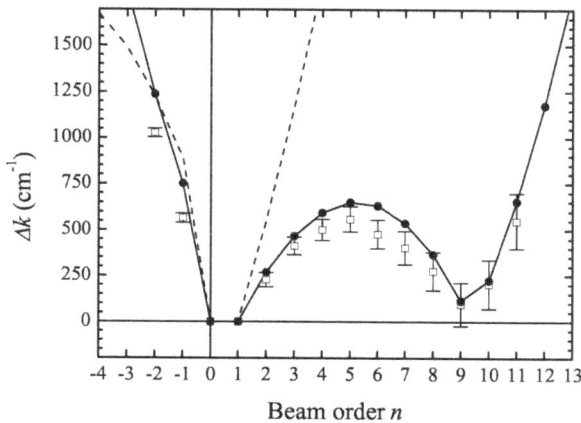

Figure 6. Wave-vector mismatch of the beams in the pattern: open squares, measured points; filled circles, calculation for the cascaded processes; dashed curve, calculation assuming collinear geometry ($\theta = 0°$). Adapted from [11].

The experiment and the simple description based on momentum and energy conservation laws already show that the process is of universal character and will occur in any material with a third order susceptibility. Hence, all materials can show CFWM processes. Indeed CFWM processes was shortly thereafter shown to occur in gases [13] and later found to occur in other materials such as sapphire [15] and in crystals [16] or using two color beams from a fundamental laser and the output of a hollow fiber [17], which also demonstrated the intrinsic tunability of the process. Highly nondegenerate CFWM was also shown to occur in an optical parametric oscillator via cascaded $\chi^{(2)}$ processes giving rise to an effective third-order nonlinearity [18]. Recently, almost octave-spanning spectra were produced by CFWM of two synchronized ps lasers in optical microfibers [19]. Tunable CFWM has also been obtained by crossing two chirped pulses with the same central wavelength and variable delay in glass [20].

3. Knowing the Fields: The Complete Theoretical Model

Although the previous reasonings based on energy and momentum conservation allow us to understand the generation of multicolored light pulses departing from two-color pumps by cascaded FWM processes, enabling the estimation of the frequencies and emission angles of the generated beams, no predictions can be made for example regarding the energy carried by each color, nor the emission bandwidth and phase of the generated fields (including the time at which each beam is generated). Hence, a more complete model was required, which was developed based on the Slowly Varying Envelope Approximation (SVEA) for the propagation of the fields in 2.5 D (propagation along a plane while assuming cylindrical symmetry in each beam) that accurately reproduces the experimental results. The use of two spatial dimensions is important to account for angular phase-matching and actual beam overlapping region.

The wave equation for an electric field $E(r,t)$ propagating in a nonlinear medium with a polarization comprising a linear and a nonlinear term as given by $P(r,t) = P_L(r,t) + P_{NL}(r,t)$, is

$$\nabla^2 E(r,t) - \frac{1}{c^2}\frac{\partial^2 E(r,t)}{\partial t^2} = \mu_0 \frac{\partial^2 P(r,t)}{\partial t^2} \tag{5}$$

with c the velocity of light and μ_0 the vacuum permeability, where we have written the electric field as a scalar quantity, which is correct for linearly polarized fields as is the case of our experiments. Considering that the electric field is well represented under the SVEA approximation, i.e., $E(r,t) = 1/2A(r,t)e^{i(k_0z-w_0t)} + c.c.$, changing the equation to the reference frame which moves with the group velocity of the pulse ($k_1 = 1/v_g$, where v_g is the group velocity of the pulse at frequency w_0), separating the polarization into its linear and nonlinear parts and Fourier transforming to the frequency domain, the wave equation becomes

$$\frac{\partial^2 A(r,\omega-\omega_0)}{\partial z^2} + 2i\left[k_0 + k_1\left(\omega-\omega_0\right)\right]\frac{\partial A(r,\omega-\omega_0)}{\partial z} +$$
$$\left[k^2\left(\omega\right) - k_0{}^2 - 2k_0 k_1\left(\omega-\omega_0\right) - k_1{}^2\left(\omega-\omega_0\right)^2 + \nabla_\perp{}^2\right]A\left(r,\omega-\omega_0\right) = \qquad (6)$$
$$-2\mu_0\omega^2 P_{NL}\left(\omega\right)e^{-ik_0 z}$$

This equation describes both linear and nonlinear propagation. It includes dispersion and diffraction, as well as a particular form of P_{NL} which includes the instantaneous non-resonant Kerr effect and self-steepening. Note that no expansion of the wave vector k was performed and therefore higher order terms of the dispersion are intrinsically included. To solve the equation, it was separated in its linear and nonlinear parts. For $P_{NL} = 0$, the linear part of Equation (6) in the paraxial approximation ($\frac{\partial^2 A(r,\omega-\omega_0)}{\partial z^2} = 0$) is given by

$$2i\left(k_0 + k_1\left(\omega-\omega_0\right)\right)\frac{\partial A(r,\omega-\omega_0)}{\partial z} +$$
$$\left[k^2\left(\omega\right) - k_0{}^2 - 2k_0 k_1\left(\omega-\omega_0\right) - k_1{}^2\left(\omega-\omega_0\right)^2 + \nabla_\perp{}^2\right]A\left(r,\omega-\omega_0\right) = 0 \qquad (7)$$

This linear part carries information about dispersion and diffraction and can be solved in the frequency domain by using a (2,2) Padé approximant for wide angle propagation.

The nonlinear part in the paraxial approximation is given by

$$2ik_0\frac{\partial A\left(r,\omega-\omega_0\right)}{\partial z} = -2\mu_0\omega^2 P_{NL}\left(\omega\right)e^{-ik_0 z} \qquad (8)$$

The nonlinear polarization P_{NL} can be written in terms of a slowly varying envelope p as $P_{NL}\left(r,t\right) = 1/2p\left(r,t\right)e^{i\left(k_0 z - w_0 t\right)} + c.c.$, with $p\left(r,t\right) = 3/8\epsilon_0\chi^{(3)}\left|A\left(r,t\right)\right|^2 A\left(r,t\right)$ and is better described in the time domain. By doing the inverse Fourier transform, we get

$$\frac{\partial A\left(r,t\right)}{\partial z} = \frac{3\chi^{(3)}}{8ik_0 c^2}\left[-\omega_0{}^2\left|A\right|^2 A - 2i\omega_0\frac{\partial\left(\left|A\right|_2 A\right)}{\partial t} + \frac{\partial^2\left(\left|A\right|^2 A\right)}{\partial t^2}\right] \qquad (9)$$

where the first term on the right hand side accounts for self-phase modulation effects and the second for self-steepening. The accurate evaluation of the self-steepening term is essential, because the spectral span of the generated CFWM pulses is of the same order as the central frequencies of the pumps, ω_0 and ω_1. It should be pointed out that no additional delayed response terms were required in order to faithfully reproduce the experimental results. This equation can be solved by a second-order Runge-Kutta method, which proved sufficiently accurate.

Equations (8) and (9) are very powerful in the context of the present work and can exhibit a very rich behavior, since they constitute a much better approach to the process than the simple algebraic description of the previous section. Let us consider the geometry of Figure 7 with spatial gaussian profiles $A(r, t)$ or $A(r, \omega - \omega_0)$ and transform limited pump pulses, for the two beams used in the first experiment: an orange beam (ω_0, λ_0 = 618 nm, 80 fs) and a green beam (ω_1, λ_1 = 561 nm, 40 fs), interacting at an internal angle of 1.93° (corresponding to an external angle of 2.9°) in a BK7 glass slide with a relative delay of 0 fs between them.

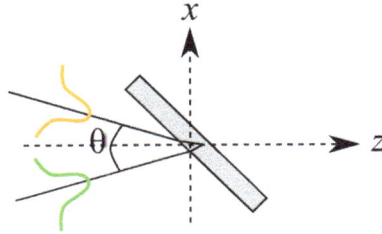

Figure 7. Interaction geometry.

The resulting simulated CFWM spectrum is shown in Figure 8 (black curve). We see that several upshifted orders are generated, which are broadband and have well-defined central wavelengths. Also, the total spectral intensity has an envelope which decays at around 400 nm and then increases again, in agreement with the experimental observations shown in the next section. If the interaction occurs at a slightly different angle, which is a feasible situation in an experimental setup, the efficiency and the generated central wavelengths change, as shown in Figure 8 (collinear pumps don't generate more than 1–2 CFWM orders since there is no phase-matching). For angles smaller than 1.93°, the generated orders don't reach so far into the UV, while for angles larger than 2° the overall efficiency drops rapidly (Figure 8a). For a fixed interaction angle, different time delays between the pump and the signal can be set, which also influences the central wavelength of the generated orders due to additional shifts caused by cross-phase modulation and the varying amount of temporal and spatial overlap between the two pump beams (Figure 8b). We see that for delays of ± 20 fs (typical reading precision of a manual delay line) around the optimum zero delay point, the overall efficiency drops by only a small amount (more noticeable for higher orders) since such delays are still well within the 40 and 80 fs FWHM pump pulse durations. For delays larger than ± 40 fs the efficiency starts to drop significantly, as illustrated in Figure 8b for the +40 fs case. The model also offers the possibility of spatially shifting the pumps at the entrance of the slide. Figure 8c shows the dependence of the generated field intensity for different spatial overlaps of the pump beams. For ± 50 μm shifts in the overlap of the pump beams (each 100 μm in diameter) at the entrance plane of the medium, the overall efficiency decreases roughly by a factor of 2 with respect to the perfect overlap (0 μm) case. This is again more noticeable for the higher orders in the ultraviolet and near-infrared regions.

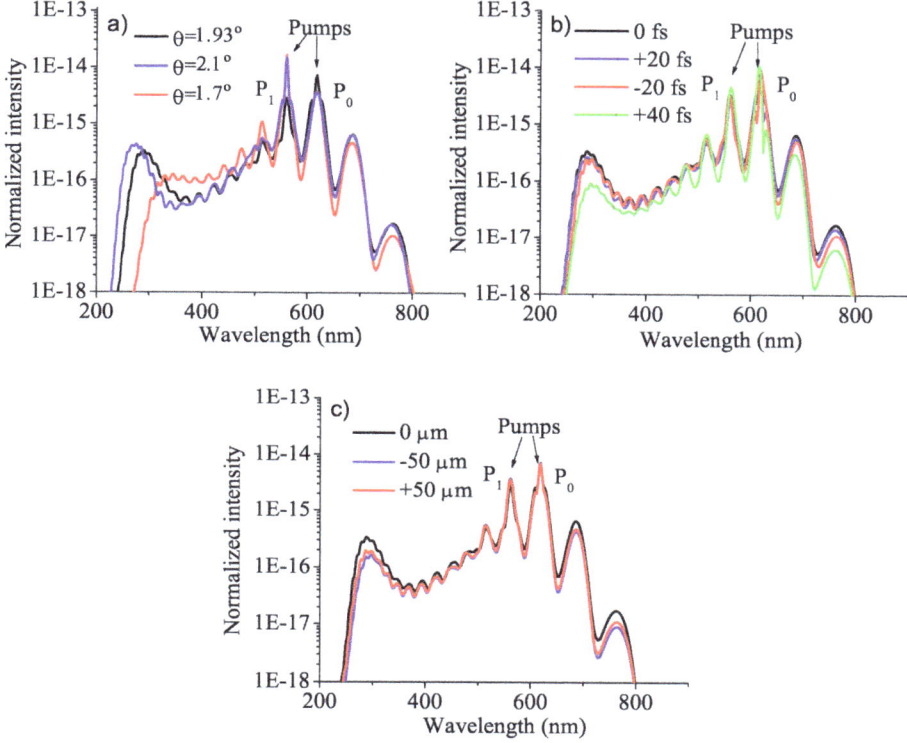

Figure 8. (**a**) Generated spectra for different interaction angles; (**b**) Generated spectra for 1.93° and different relative time delays between pumps (positive delay P_1 arrives before P_0, negative delay P_0 arrives before P_1); (**c**) Generated spectra for θ = 1.93°, 0 fs relative time delay and different entrance x-positions at the slide.

Hence it is possible to tune the generated spectra by varying the interaction angle and/or the delay, in very good agreement with experimental observations. This means that the terms considered in the linear and nonlinear sub-equations are sufficient to faithfully reproduce the experimental results, as seen further below. The experiment was repeated in a 150 μm fused silica slide to obtain more upconverted orders in the UV and the internal interaction angle was 1.57°. In this situation, multiple CFWM orders were generated up to the 21st upshifted order at 209 nm (made visible with the help of a fluorescent card) and were registered with a broadband fiber-coupled spectrometer, but due to spectral limitations the measurement reached only down to 250 nm. The temporal characteristics of the beams were measured by polarization-gating FROG (PG-XFROG). More details about these experiments are given in the next section.

The simulations provide the time dependent field distribution at the end of the glass slide as a function of the transverse coordinate x and frequency ω. The corresponding total spectrum (obtained by integrating along the transverse direction) for an interaction angle of 1.57° is given in Figure 9 and clearly shows the asymmetric generation of up and downconverted orders. The efficiency of the generated frequency upshifted pulses follows the same pattern as the experiment, although it has a sharper cutoff after the 12th order that depends on the interaction angle and pulse energy. This discrepancy is possibly due to experimental intensity calibration errors, which are larger at the spectral edges.

A look at the fields inside the slide shows that all orders are practically generated within the first 60 μm of material. The spatial resolution of the model allows determining at which angle each of the orders is generated. This can be seen in the $\theta - \lambda$ plot in Figure 10.

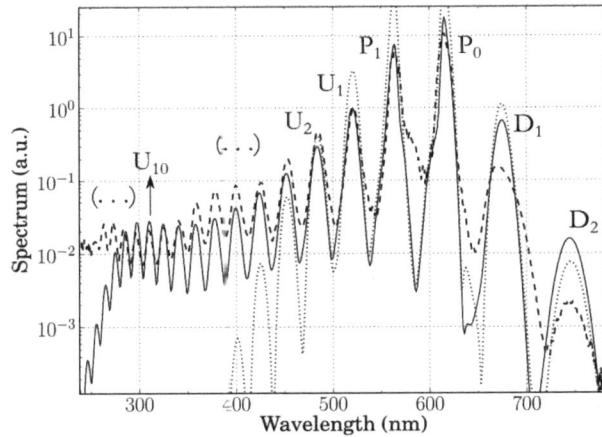

Figure 9. Spectrum of the multiple CFWM orders generated in fused silica spanning over 1.5 octaves (the two pump frequencies correspond to the two larger peaks). Dashed line: Experimental results (see [21]), Solid line: Numerical simulation, Dotted line: Numerical simulation for collinear pumps [22].

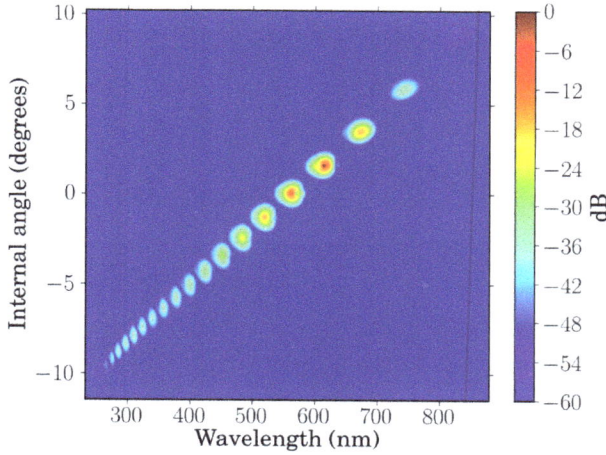

Figure 10. Simulated θ − λ spectrum of the CFWM beams, clipped at 10^{-6} of the maximum [22].

The temporal characteristics of the pulses can be also calculated, since the model gives us complete temporal and spectral information of the field. A given CFWM order can be spatially selected since they are emitted at different angles and we can simulate its interaction with another beam, as done in experiments for measuring the temporal duration. For instance, one can make a given generated order interact with the beam at ω_0 on a second thin slide of fused silica for different relative temporal delays and hence numerically obtain polarization-gating cross-correlated frequency resolved optical gating (PG-XFROG) traces. Figure 11 shows the measured and simulated PG-XFROG traces for the second frequency upshifted beam with central frequency ω_2 (top) and the first downshifted beam with central frequency ω_{-1} (bottom). I, the simulation results are in good agreement with the measurements regarding the broad bandwidth and short duration of the newly generated low-order pulses (which are shorter than the pumps). There are however differences in the central frequency and in the chirp of the pulses. The former can be due to several factors, from changes in the central frequency of the pumps that can occur between experiments, to differences in the interaction angle as well as in the relative pulse delay, all of which can result in frequency shifts, as described previously (see, e.g., Figure 8). Regarding the latter, the simulated pulses exhibit some positive chirp due to the nonlinear phase imposed by the pumps. This is clearer in the right plot of Figure 12, which shows the calculated PG-XFROG traces for the synthesized field of Figure 13. Another factor that directly contributes to the chirp of the newly generated pulses is the initial chirp of the pump pulses. In the simulations the pumps are assumed to be Fourier-limited, whereas in the experiments there can be slight deviations to this condition.

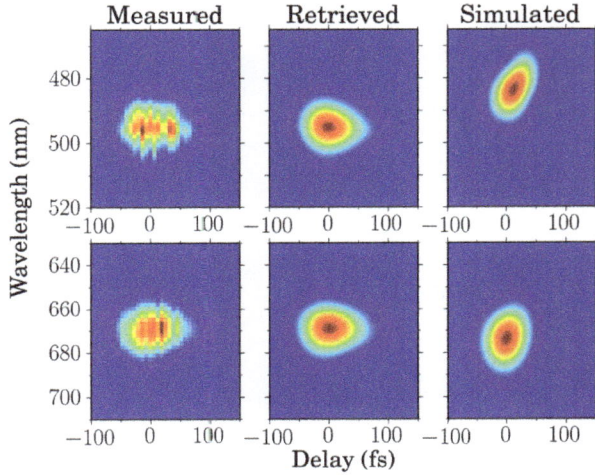

Figure 11. Experimental (measured and retrieved) and simulated polarization-gating FROG (PG-XFROG) traces of the second frequency-upshifted pulse (top row) and first downshifted pulse (bottom row). The retrieved pulse durations are 30.7 and 40.2 fs for the second upshift and first downshift respectively (FROG error was 0.014 and 0.008 for a 128×128 grid) [22].

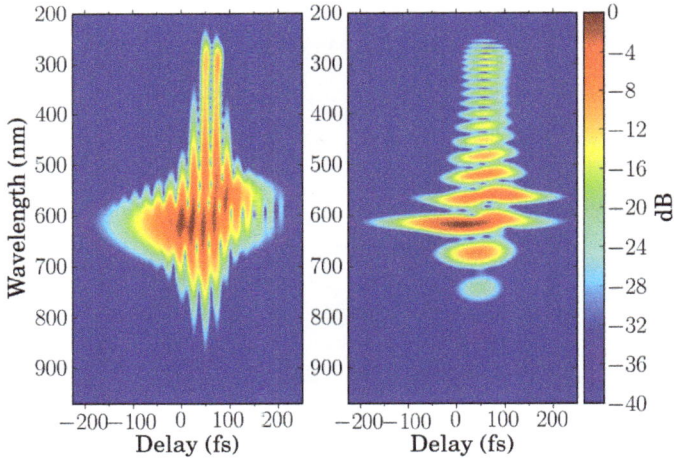

Figure 12. Simulated PG-XFROG, clipped at 10^{-4} of the maximum, for the recombined and focused CFWM field, assuming 20 fs (**left**) and 80 fs (**right**) gate pulses [22].

48

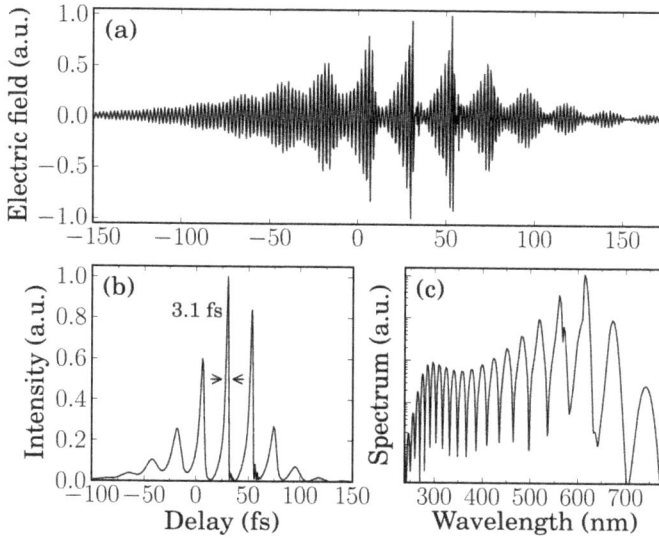

Figure 13. Recombined and focused CFWM pulses. (**a**) Electric field as a function of the delay (in fs); (**b**) Intensity; (**c**) Power spectrum (logarithmic scale) [22].

Other kinds of temporal processing are also possible. For example the total electric field can be calculated by synchronizing and spatially superimposing all the CFWM orders. Since after passing the slide the different orders have acquired a slightly different group delays, synchronization can be achieved by subtracting the linear phase. This is equivalent to focusing all the orders together in a point, which would be an experimental approach for having the total field in the same region. The total electric field obtained by this numerical procedure can be seen in Figure 12a, along with the field intensity in Figure 12b and its power spectrum in Figure 12c. We see that in the time domain we obtain a train of pulses separated by the pump beat period (23 fs) with the central peak having 3.1 fs in duration.

A PG-XFROG trace of this total field can also be simulated. The gate pulse can be of course arbitrarily chosen, but it looks reasonable and informative to use one of the pump pulses as gating pulse, as done in the experiments, namely the 80 fs pulse at ω_0. The PG-XFROG trace for the complete focused set of CFWM orders is shown in Figure 13 where the different sub-traces for the individual orders can be distinguished (on the right). For shorter gate pulses spectral resolution is lost but temporal resolution is improved revealing the fine temporal structure (pulse train) obtained in this case (on the left).

4. Towards Single-Cycle Pulse Synthesis

The described set of cascaded FWM orders will be a coherent broadband source provided the different orders can be spatially and temporally overlapped. Coherent spectra that are sufficiently broad can be used for generating few- and single-cycle pulses if properly compressed. A well-known technique for generating high energy few-cycle pulses is the temporal compression of the broadband supercontinuum produced by self-phase modulation (SPM) in e.g., gas-filled hollow core fibers [23] (discrete spectra, composed of evenly spaced near-monochromatic waves can also lead to the synthesis of a train of single-cycle optical pulses, as demonstrated using molecular modulation in a gas driven by two independent nanosecond lasers [24]). Moreover, the central frequency of the total CFWM spectrum and the separation between sidebands can be freely adjusted by tuning the pump frequencies, so in principle it should always be possible to obtain the commensurate sidebands required to synthesize a train of identical pulses. Two frequencies are commesurate if $m\omega_0 = n\omega_1$, with m, n integers. This also means that each frequency is an exact multiple of the frequency spacing $\omega_m = \omega_1 - \omega_0$.

Let us take a deeper insight into pulse synthesis by addition of a finite number of pulses with different central wavelengths. The pulses must have fixed relative phases in order to produce a stable coherent superposition of fields giving rise to an ultrashort pulse or to a train of ultrashort pulses. We also found this result in the previous section (Figure 12), where the coherent sum and focusing of the CFWM orders generated by numerically solving the equations of our model, resulted in a train of ultrashort laser pulses. For the CFWM pulses this can be understood using a simplified model based on the coupled amplitude equations between the pumps and the generated sidebands.

Using the wave equation in the slowly-varying envelope approximation, for two pump fields with complex amplitudes $\tilde{E}_0 = E_0 e^{i\phi_0}$ and $\tilde{E}_1 = E_1 e^{i\phi_1}$, frequencies ω_0 and ω_1, respectively, assuming a cubic nonlinearity, neglecting dispersion and pump depletion, and assuming perfect phase-matching and constant coupling coefficients, the solution for the complex amplitude $\tilde{E}_n$ of the field with frequency ω_n at the exit of a medium with length L can be written as [21,25]

$$\tilde{E}_n(s) = i^n e^{in\delta} \tilde{E}_0 J_n(s) - i^{n+1} e^{i(n-1)\delta} \tilde{E}_1 J_{n-1}(s) \tag{10}$$

where $s = \gamma L$ is the nonlinear phase shift acquired in the medium, $\delta = \phi_1 - \phi_0$ is the initial phase difference between pumps, J_n are the nth order Bessel functions of the first kind, and the nonlinear coefficient is given by $\gamma = 3\omega_0 \chi^{(3)} E_0 E_1 / (2n_0 c)$, with n_0 the linear refractive index at frequency ω_0 and c the speed of light in vacuum. Equation (1) predicts well-defined phase relationships between the several CFWM orders, thus establishing mutual coherence between the beams. The total field in the time domain is then

$$E(t) = \text{Re} \left[\sum_n \tilde{E}_n(s) \, e^{-i(\omega_0 + n\omega_m)t} \right] \tag{11}$$

From Equation (11) and using the Jacobi-Anger identity, $\sum_n J_n(z) \exp(in\theta) \, i^n = \exp(iz\cos\theta)$, we obtain

$$E(t) = Re \left[e^{i\phi_0} \left(E_0 e^{-i\omega_0 t} + E_1 e^{i\delta} e^{-i\omega_1 t} \right) \right. \\ \left. \times e^{-is\cos(\omega_m t - \delta)} \right] \tag{12}$$

This result is similar to the frequency-modulation (FM) solution by Harris [24] for coherent addition of Raman sidebands in the approximation of negligible dispersion and limited modulation bandwidth (on its hand similar to the solution previously obtained by Lichtman *et al.* [6] for nondegenerate FWM of 100 ns pulses - for which dispersion was practically negligible - propagating in 400 m long optical fibers), with the important difference that in CFWM the synthesized pulse envelope strongly depends on the initial phase difference δ. In molecular modulation, the detuning between the driving frequency and the Raman transition dictates the preparation of a phased or an anti-phased molecular state, the latter resulting in a negative effective nonlinear coefficient γ for which the synthesized pulses could have an approximately negative chirp and hence could be further compressed by simple propagation in a normally dispersive medium [24]. In the present case of CFWM, the sign of γ is independent of the pump field (and usually positive for most optically transparent $\chi^{(3)}$ media), but the chirp of the synthesized pulses can nevertheless be controlled by adjusting δ alone. From Equation (12) we see that ϕ_0 only affects the carrier envelope phase (CEP) for the pulse train, defined with respect to the two-color beat envelope, whereas δ is associated with net temporal shifts and also determines the pulse chirp, as given by the oscillating nonlinear phase term. In particular, when $\delta = \pm\pi/2$, the synthesized pulses can have a negative chirp. In practice, a given phase difference can be introduced by slightly adjusting the delay between the two (phase-locked) pump pulses without significantly affecting their temporal overlap. Also, if the pump frequencies are commensurate, this will result in the generation of identical pulses that are CEP-stabilized within each train (although their CEP may change from shot-to-shot), since even though each CFWM step is not self-CEP-stabilized, the total recombined field will be. The corresponding pulse envelopes will also be identical from shot-to-shot, provided that the relative phase difference δ remains constant, which is a far less stringent requirement than the need of CEP stabilized pump pulses.

To demonstrate the feasibility of this approach for the generation of single cycle pulses, the basic set-up of Figure 2 was improved by using fused-silica instead of BK7 to generate more orders in the UV, and complemented with additional optics for pulse synthesis and measurement. This set-up is reproduced in Figure 14-I. As in the previous case, two-horizontally polarized visible femtosecond pulses from a dye laser-amplifier at 10 Hz were used as pumps: an orange pump with $\lambda_0 = 615$ nm, ~80 fs duration, 2 mJ energy and a green pump with $\lambda_1 = 569$ nm, ~60 fs duration and 200 µJ energy. Special care was taken in using Glan-Taylor polarizers (GTPs, 100,000:1 extinction ratio) to have perfectly polarized beams, which is important not only for the generation, but for the subsequent temporal measurements. In this laser system, the green pump beam is directly (optically) derived from the orange pump beam: a small portion of the orange beam is first used to generate supercontinuum in a cell filled with deuterated water, and then the green portion of this supercontinuum is amplified using green laser dyes. Even though supercontinuum generation and laser action are coherent processes, the relative phase between the two pulses is not actively locked and can fluctuate due to normal thermal and mechanical perturbations and drifts in the system. The pulses are commensurate to within 0.2% (well within their ~5 nm bandwidths), hence we can expect the resulting synthesized pulses to have the same CEP within the pulse train.

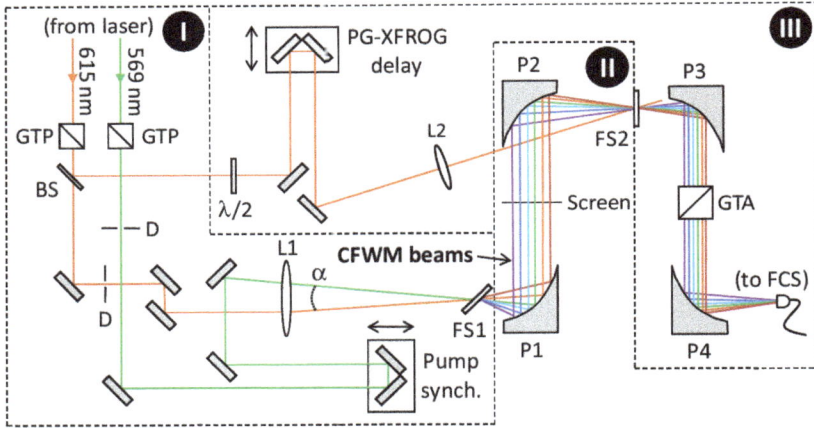

Figure 14. Experimental setup: (I) generation of CFWM pulses; (II) pulse recombination and synthesis; (III) pulse characterization by PG-XFROG (see text for details) [21].

The orange and green pulses are synchronized at the entrance plane of the fused silica slide FS1, and have energies of 32 and 38 µJ, respectively, near-transform-limited durations (measured using SHG-FROG), beam radii of approximately 80 and 100 µm, respectively, and similar intensities on the order of 2×10^{12} W/cm^2 as they cross on the slide at an angle of $\approx 3°$. Under these conditions, a fan of 20 upconverted CFWM orders was generated up to 209 nm, as shown in Figure 15a. The total measured energy in the cascaded beams (excluding the pumps, *i.e.*, orders 0 and 1) is > 6 µJ, so approximately 10% of the incident energy is transferred to the newly generated frequencies.

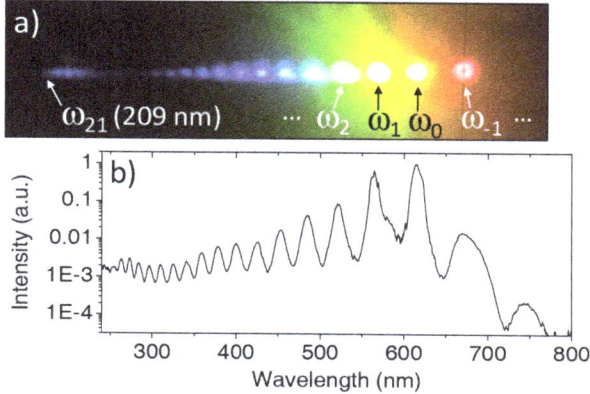

Figure 15. (a) Direct (no filtering) image of the fan of multicolored CFWM pulses as seen projected on a phosphor-coated paper screen (the dark arrows denote the pump beams); (b) Corresponding two-octave spectrum measured at the focal plane of mirror P2. [21].

To perform pulse synthesis all orders were collimated and recombined in a single 100 µm white light spot using two $\lambda/8$ Aluminum-coated off-axis parabolic mirrors P1 (f = 2.54 cm) and P2 (f = 5.08 cm) (Figure 14-II). A perforated screen was placed between these two mirrors to transmit only the central portions of the pumps and thus have a more balanced spectrum, but no additional amplitude filtering was performed. The total spectrum at the focal spot was measured with an intensity calibrated UV-NIR (200–1100 nm) spectrometer, but only 15 frequency upconverted orders could be detected due to the limited bandwidth of the mirrors, as shown in Figure 15b. All orders have been generated under simultaneous phase-matching conditions with a fast electronic nonlinearity, so they are phase-locked when exiting the first fused silica slide FS1. Additional propagation in air only adds negligible second-order dispersion and hence we expect a train of few-cycle pulses to be synthesized at the focus of P2.

Temporal characterization was done with polarization gating XFROG (PG-XFROG) using an 80 fs pulse as the gate (Figure 14-III). This gating pulse was obtained from the main orange beam with a 50/50 beamsplitter (BS) and its polarization was rotated 45° with a half-wave plate. This pulse induces birefringence in another fused silica plate (FS2) placed at the focal plane of mirror P2, where the pulse synthesis using all CFWM orders is taking place. A second Glan-Thompson polarizer was used as analyzer (GTA) crossed with the original GTPs, which required collimating the beams with a third parabolic mirror P3 (f = 2.54 cm). Finally the total spectrum of the gated pulses was measured by focusing the output of the GTA with mirror P4 (f = 5.08 cm) onto an optical fiber with a large 400-μm core coupled to a spectrometer (FCS). The PG-XFROG trace obtained by registering the spectrum of the gated pulses as a function of the delay of the gating pulse can be seen in Figure 16a, where up to 13 orders could be simultaneously gated, so they were synchronized, and also practically have no chirp. Since the gate pulse was long compared to the expected pulse train structure, information about the temporal structure of the traces is lost and retrieval of the whole field with this trace is not reliable (see previous section where a simulated PG-XFROG trace using a 20 fs gate pulse showed the fine temporal structure). However, unambiguous retrieval of the individual orders is possible. Figure 16b,c show the measured and retrieved PG-XFROG traces of the first upconverted CFWM pulse and its associated temporal shape with 30.6 fs in duration. Pulse compression and synthesis of the generated CFWM beams have also been studied and reported by the T. Kobayashi (see, e.g., [26]) and A. H. Kung [14] research groups.

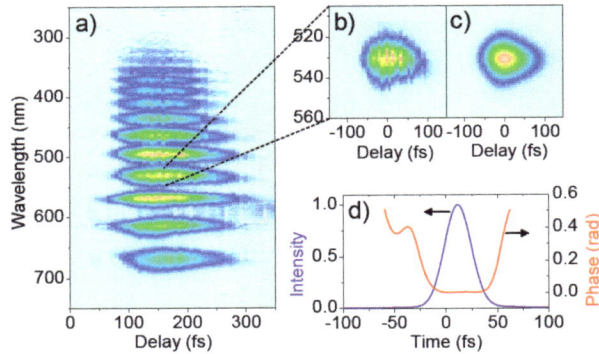

Figure 16. (a) Measured PG-XFROG trace of the synthesized field; (b) Measured and (c) retrieved PG-XFROG traces of the first frequency upconverted CFWM pulse, and (d) corresponding intensity and phase in the time domain. [21].

Adding the fields obtained for all measured orders while setting zero relative delay between each CFWM pulse, a synthesized field is obtained. This field has a main central transform-limited pulse with 2.2 fs duration and two smaller pulses, one at each side on the main pulse, separated 25 fs from the main pulse, which corresponds to the pump beat period. Figure 17 shows the synthesized field along with the pulse intensity and evidences the possibility of synthesizing single cycle pulses with this technique. The central pulse carries almost all the energy, which amounts to 5 μJ.

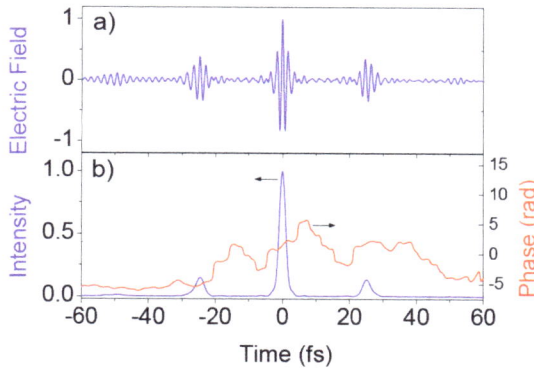

Figure 17. Total field (**a**) Normalized electric field and (**b**) intensity and phase of the pulses obtained by coherent addition of the retrieved electric fields of the 13 gated CFWM pulses. The main peak is a 1.3-cycle, transform-limited 2.2 fs pulse. [21].

5. Generation of Ultraviolet (UV), Deep Ultraviolet (DUV), Vacuum Ultraviolet (VUV) and Higher-Order Harmonics by CFWM from a Standard Titanium:Sapphire Laser Amplifier

Since the materials have usually resonances in the UV, the use of pumps in the visible part of the spectrum means working on the normal dispersion region. Angle phase-matching of upconverted orders in CFWM processes is possible thanks to dispersion. The shorter the wavelength of the resonance in the material the higher the number of upconverted orders that can be obtained, provided that the form of the dispersion curve is adequate. In the previous results we showed how cascaded upconverted orders comfortably reached the UV and DUV up to 209 nm departing from orange and green pump pulses from a dye laser-amplifier. A natural extension of the research in CFWM is trying to reach the UV, DUV, VUV or to generate high order harmonics with more common pump colors, available in many labs, such as the fundamental and the second-harmonic of a Titanium:Sapphire laser.

5.1. Ultrashort Pulses from the UV to the VUV by CWFM

Materials with large band gap such as alkali metal halide crystals are transparent in the VUV, so they are candidates for generating a cascade of upconverted orders via FWM up to the VUV, and we have tested this numerically [27]. The model developed in Section 3 assumes isotropic materials, and some alkali metal halides have cubic symmetry, so they are equivalent to an isotropic material and we used our model to

55

test the feasibility of obtaining VUV pulses via CFWM in LiF using a pump beam at 400 nm and a signal beam at 800 nm. Equation (3) gives the internal crossing angle with would result in perfect phase-matching of the nth order, but for pump and signal beams as separated as a fundamental wavelength and its second harmonic, the internal angles for successive upconverted orders is too different for a cascade to be generated departing from only two beams, so the geometry of Figure 18 was devised, with a pump field at 400 nm and 3 signal beams at 800 nm, the latter satisfying the phase-matching condition for different consecutive orders.

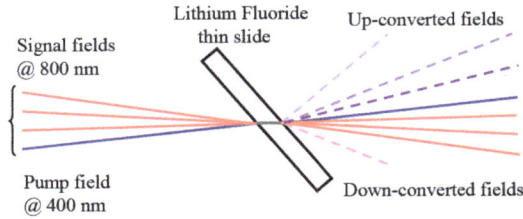

Figure 18. Scheme for multiple-beam CFWM with strong angular separation between pump and signal frequencies. [27].

We consider a pump beam at 400 nm and three signal beams at 800 nm, all transform-limited with 25 fs duration and focused to a 40 μm radius spot. All pump and signal beams have an irradiance of 5×10^{12} Wcm^{-2} (energy per pulse of ~ 6 μJ). The pump and the three signal beams form internal angles of 11.98, 15.65 and 20.06 degrees in a 300 μm thick LiF slide. The signal beams can be individually delayed to optimize the VUV generation. Figure 19 shows the CFWM spectrum generated for a time delay of 0 fs or 7 fs (for all signal beams) along with the spectrum generated by adding the individual spectra obtained by using the pump pulse and one of the signal beams at a time, also with a time delay of 7 fs.

In these conditions, we observe generation of CFWM beams up to the 6th harmonic of 800 nm, with good efficiency up to the 5th harmonic. Only a slight dependence on the time delay is seen and there is full evidence of the existence of CFWM processes, since the VUV spectrum obtained by addition of the individual spectra from the different pairs of pump and signal pulses is much weaker. As already seen in section 3 the generated orders are angularly separated and due to the fact that all three pump beams at 800 nm contribute to generate a specific order at a slightly different interaction angle and wavelength, because of the different phase-matching conditions, the generated orders have a complex spatial structure, although angular selection of a specific order can always be performed. Figure 20 shows the simulated $\theta - \lambda$ spectrum for 7 fs and 0 fs delay. A cleaner structure and also a better efficiency in the 4th and 5th harmonic is observed in the case of 7 fs delay compared to perfect synchronization (0 fs delay).

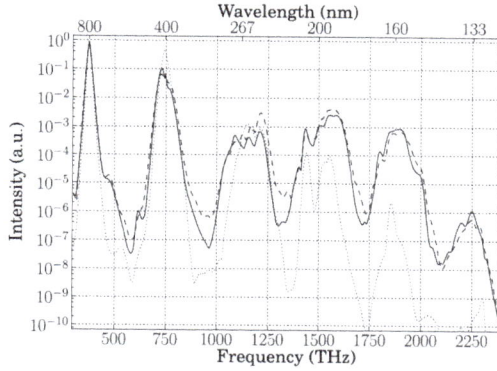

Figure 19. Simulated CFWM spectra in LiF. Solid (dashed) curve: spectrum generated by the interaction of all three signal pulses at 800 nm delayed by 7 fs (0 fs) with respect to the pump pulse at 400 nm – (see text); dotted curve: combined spectrum for three independent simulations, each optimizing an individual consecutive harmonic generated by each pair of noncollinear pump and 7 fs delayed signal pulses [27].

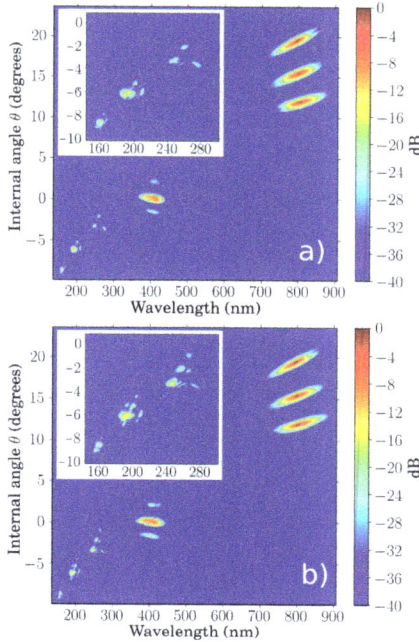

Figure 20. θ − λ spectrum for 7 fs (a) and 0 fs (b) relative delays between signal and pump pulses. Insets: magnified θ − λ spectra for the first three upconverted orders [27].

The temporal behavior is also as expected, with the generated orders having ultrashort durations of the order of the duration of the pump and signal pulses (31.9, 30.5, 30.1 and 20.4 fs for the third to the sixth harmonic respectively), due to the corresponding nonlinear spectral phases. This can be seen in Figure 21. The spectrum of all orders is broadband and assuming flat spectral phases the transform-limited durations are much shorter than those of the pump and signal pulses, namely 8.86, 8.1, 6.0 and 4.0 fs, respectively.

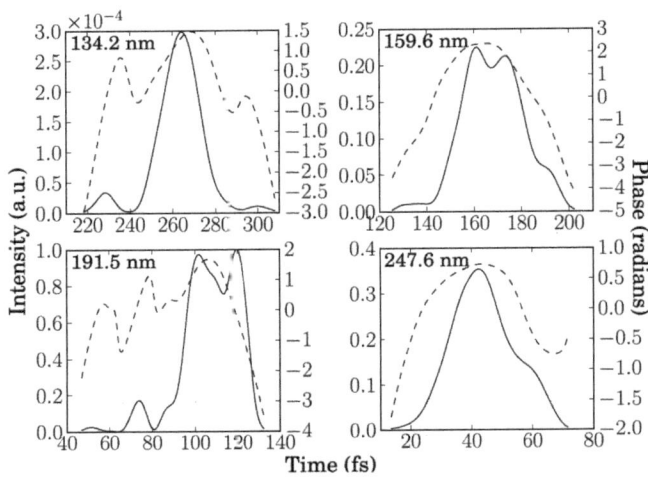

Figure 21. Temporal characteristics of the generated harmonics. Solid curve: intensity; dashed curve: phase [27].

5.2. DUV Ultrashort Pulse Generation by Highly Nondegenerate FWM

The observed fact that the generated orders are broadband deserves more attention, since the availability of sub-300 nm DUV ultrashort pulses with energies in the range of microjoules is important to study processes in photophysics, photochemistry and photobiology. Hence, although not a cascaded process, we also concentrated on studying the generation of the third harmonic of 800 nm via highly nondegenerate FWM [28]. Several techniques based on FWM in gaseous media have been employed by other groups, from third-order difference-frequency generation (DFG) of Ti:Sapphire laser pulses and their second-harmonic in hollow waveguides [3,29] and in filaments [30,31], to DFG of pre-chirped pulses in dual hollow-fiber systems [32]. Direct harmonic upconversion of few-cycle femtosecond pulses in gases [33,34] has also been performed. This last method has enabled the generation of the shortest (sub-3-fs) DUV pulses to date [35].

The setup used for the generation of broadband pulses at 266 nm by mixing two ultrashort pulses at 800 nm and 400 nm respectively in an isotropic solid is shown in

Figure 22, along with the part for the temporal measurement. A Titanium:Sapphire amplifier provides the fundamental signal (or idler) pulses at 800 nm (ω) with 27 fs pulse duration, 1 mJ energy, horizontally polarized and at 1 kHz repetition rate. The pump pulses at 400 nm (2ω) are generated in a type-I 200 μm thick BBO crystal. To maximize the SHG, the chirp of the fundamental pulse was optimized and a refractive telescope reduced its beam size to 5 mm. A dichroic mirror separated the pump and signal beams and a half wave plate HWP was used to rotate the polarization plane of the signal beam by 90° so as to have the polarization parallel to the pump beam, which emerges perpendicularly polarized from the BBO crystal. Both pump and signal beams were focused with lenses to interact on a fused silica slide FS at an internal angle of 15.75° as given by Equation (3) (external angle θ = 23.2°) for the FWM process $3\omega = 2 \times (2\omega) - \omega$. The slide was placed 2 cm before the focus of the lenses to avoid damage. Synchronization is achieved with a delay line in the signal branch of the setup and in the plane of the slide the signal and pump beams had 314 μJ and 191 μJ with estimated intensities of 2.6×10^{12} and 1.2×10^{12} W/cm^2, respectively. Under these conditions we obtained 5–6 μJ DUV pulses emitted at around 266 nm at an external angle of 7°.

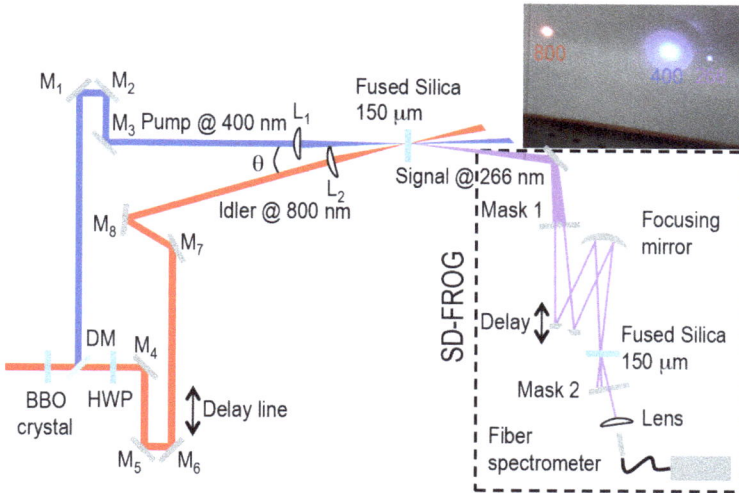

Figure 22. Experimental setup for ultrashort deep ultraviolet (DUV) pulse generation by highly nondegenerate four-wave mixing (FWM) and temporal characterization via SD-FROG. DM: Dichroic mirror, HWP: Half-wave plate; M1-M3: Aluminum mirrors; M4-M8: 45° dielectric mirrors. Inset: idler, pump and generated DUV beam, directly projected onto a white card [28].

Figure 23 shows the spectra registered with a fiber coupled spectrometer for different delays, defining the 0 fs delay for the situation in which higher energy pulses were obtained (6 µJ). Different broadband spectra were obtained for delays ranging from −99 fs to 99 fs and the dependence of the shape and central frequency can be explained by the cross-phase modulation between the pump and idler pulses. Integration of these spectra with reference to the energy measured for the one with 6 µJ (Figure 20, delay 0 fs) gives 1.5, 2.4, 3.85, 6.0, 4.75, 4.9 and 1.4 µJ for this series.

The spectra are broad enough (Figures 23 and 24) to support ultrashort durations and the DUV (as well as the SHG pump pulses) were characterized by self-diffraction FROG (SD-FROG). The experimental details can be seen in Figure 22. Two holes with 1.6 mm diameter separated by 3.5 mm were drilled on a metallic mask and aligned so that both holes transmitted similar intensities from the central part of the DUV beam. A relative temporal delay was set between both pulses using d-shaped mirrors and a stepper motor stage (3.33 fs step). Both pulses were focused with an aluminum mirror (f = 200 mm) in a second fused silica slide were a self-diffraction beam was generated and later isolated to measure its spectrum in a fiber coupled spectrometer. Figure 25 shows the measured FROG traces and corresponding retrievals using standard Femtosoft FROG software, which gives 27.3 fs DUV pulses with a spectral width of 5.7 nm and a time-bandwidth-product of 0.67. The FROG measurement of the SHG pulses resulted in a duration of 48.6 fs.

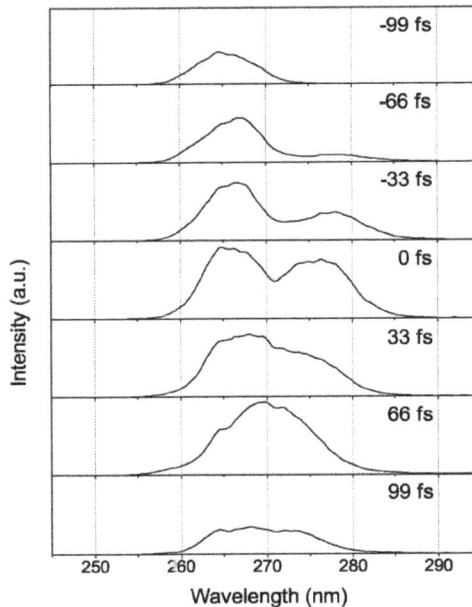

Figure 23. Experimental spectra of the generated DUV signal for different delays (separated by ∼ 33 fs) between idler and pump pulses. Negative delay: pump arrives after idler, positive delay: pump arrives before idler [28].

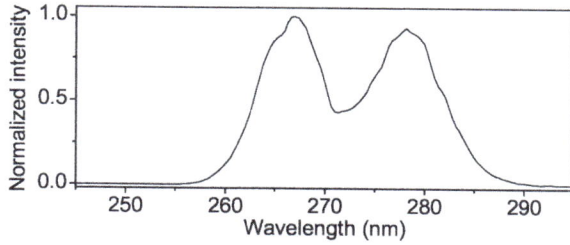

Figure 24. Broadest experimental spectrum obtained in our-set-up, capable of supporting sub-4-fs [28].

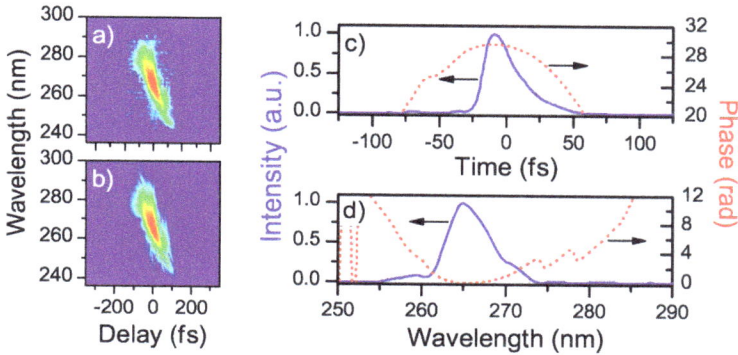

Figure 25. (a) Experimental and (b) retrieved SD-FROG traces of the DUV pulses (in log scale); (c) retrieved pulse in the time and (d) frequency domains [28].

The model described in Section 3 also predicts broadband generation of the third-harmonic of 800 nm at a central wavelength of 266 nm. Using the experimental parameters given in this section and for different internal angles θ = 15.0°, 15.7° (the angle of perfect phase-matching) and 16.4° different broadband spectra were obtained for different delays between pump and signal pulses. In Figure 26 (θ = 15.0°) we can see double band spectral structures which are enhanced in its blue portion for negative delays and in its redder portion for positive delays. The two situations are of course not symmetrical because of the different group velocities of the pulses, where the idler can catch up with the pump in situations of positive delay but will always come before the pump for negative delays. We also see that the different pulses have chirp and the more intense ones, generated at delays of −20 fs, 0 fs and 20 fs, have a clear positive chirp, mainly due to self- and cross-phase modulation. This positive chirp is in agreement with the experimental results of Figure 23, where the retrieved pulse shows also positive chirp. The calculated emission angle for the 266 nm pulse obtained for 0 fs relative delay between pump and idler was 7.3°, with an output

61

energy efficiency of 3.6% with respect to the pump pulse and a pulse width of 19.7 fs (9 fs TL duration), all in reasonable agreement with the experimental results.

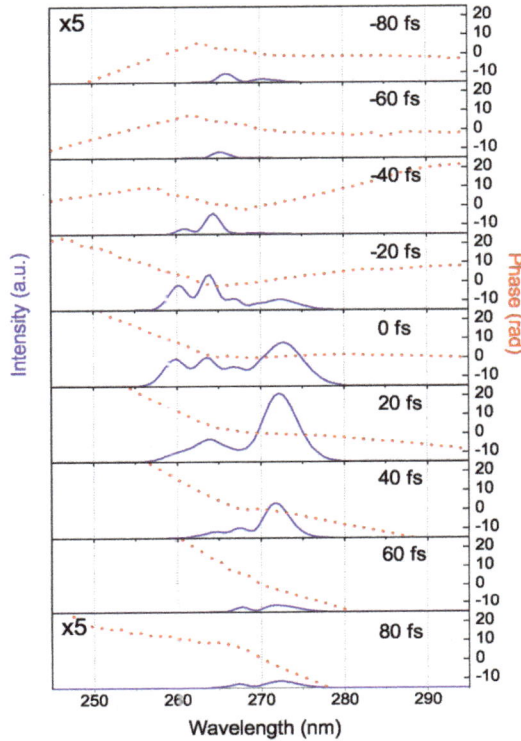

Figure 26. Simulated spectral intensity and phase of the generated DUV signal for θ =15.0° and different delays between idler and pump pulses (The intensities in the first and the last plots have been multiplied by a factor of 5). Negative delay: idler arrives before pump, positive delay: idler arrives after pump [28].

For other interaction angles (θ = 15.7° (Figure 27) and θ = 16.4° (Figure 28)) higher efficiencies are obtained (12.5% and 6.9% respectively) but with narrower bandwidths. The situation with θ = 15.0° seems to better represent the experimental one. We see that the change in interaction angle from 15 to 16.4 ° also produces a significant shift/tuning of the central frequency of the generated DUV pulses. This is due to the large angular sensitivity of the geometric phase matching condition for the highly nondegenerate FWM process involving ω and 2ω beams crossing at a very large angle.

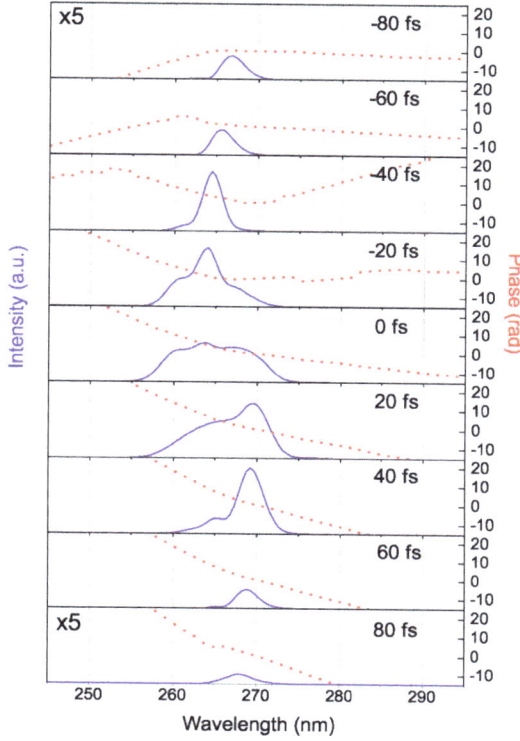

Figure 27. Simulated spectral intensity and phase of the generated DUV signal for θ = 15.7° and different delays between idler and pump pulses. (negative and positive delays as in Figure 23; note the scale factors for the first and last intensity plots) [28].

Given the rather large angles used between the interacting beams, the generated pulse has angular chirp. Figure 29 shows the angle at which each frequency is generated for the 0 fs delay of Figure 26. We see that the spectrum spans from 255 nm to 280 nm only within 0.5°, which is small enough for many applications, namely those requiring focused pulses.

Many laboratories nowadays have Titanium:Sapphire amplifiers with pulses longer than those used here. Our model predicts that one can still obtain 42 fs DUV pulses when using 100 fs pump pulses interacting at 15°, with 0 fs relative delay and with the same intensities used in the experiment and the previous simulations, $I_\omega = 2.6 \times 10^{12}$ and $I_{2\omega} = 1.2 \times 10^{12}$ W/cm^2 respectively. For $I_\omega = 2.0 \times 10^{12}$ and $I_{2\omega} = 2.0 \times 10^{12}$ W/cm^2 29.5 fs DUV pulses were obtained. These results show that it is possible to obtain sub-30 fs pulses starting from 100 fs pump and idler pulses, which might be of great interest in many labs.

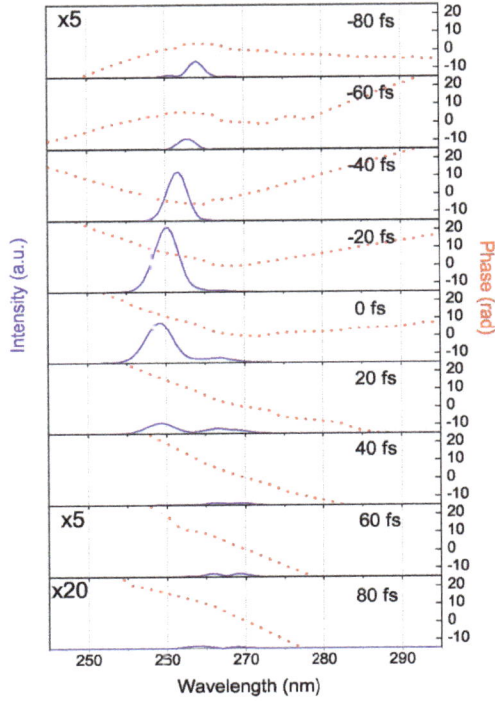

Figure 28. Simulated spectral intensity and phase of the generated DUV signal for θ = 16.4° and different delays between idler and pump pulses. (negative and positive delays as in Figure 23; note the scale factors for the first and last two intensity plots) [28].

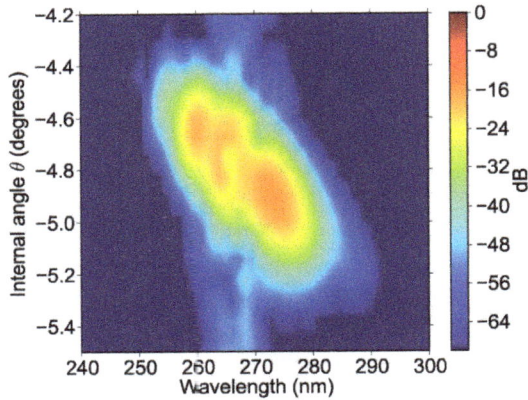

Figure 29. Angular chirp of the generated beam for the case of 0 fs delay in Figure 5. The intensity is given in dB with respect to the maximum intensity of the pump pulses) [28].

5.3. Generation of Higher-Order Harmonics by CFWM

The generation of even higher-order harmonics via CFWM, by using a fundamental beam at 800 nm at its second harmonic at 400 nm, was suggested and studied numerically shortly after the first experimental demonstration of highly-nondegenerate CFWM by the same authors [25], where the simplified analytical model further developed in [21] was also presented. Unlike usual non-perturbative high-harmonic generation in gases, which involves ionization and recombination of an electron wave-packet in the presence of an intense laser field, the proposal in [25] was based on a simple perturbative model of CFWM. More recently, Bertrand *et al.* demonstrated that noncollinear high-harmonic generation of fundamental and second-harmonic ultrashort laser pulses can be fully understood in terms of perturbative nonlinear optical wave mixing [36].

6. Conclusions

In this paper we presented an overview of our work in ultrafast highly-nondegenerate cascaded four-wave mixing in bulk media, with emphasis on key experimental results as well as on a detailed theoretical model of the phenomenon that can be solved numerically. We described how to perform an experiment where, thanks to the instantaneous third-order nonlinearity of a thin isotropic transparent medium, two noncollinear femtosecond pump pulses with different colors can give rise to a fan of broadband multicolored beams (both frequency upconverted and downconverted) via cascaded four-wave mixing, with macroscopic efficiencies in the range of 5%–10%. The newly generated beams have broad bandwidths, extending from the infrared to the ultraviolet, with each beam corresponding to a particular order within the cascade. A simple algebraic model based on energy and momentum conservation laws enables calculating the optimum interaction angle for generating a particular order n, as well as estimating the frequencies of the newly generated beams and their corresponding emission angles.

Other important characteristics such as pulse energy, bandwidth, duration or generation efficiency have been calculated and compared with the experimental results by using a unique 2.5-D nonlinear propagation model developed in our group, which takes into consideration the effects of dispersion, diffraction, self-phase modulation and self-steepening in the SVEA approximation. With this model we have been able to faithfully reproduce all the experiments performed so far. We then showed how the fan of beams produced by cascaded four-wave mixing in thin low-dispersion media, which are mutually coherent and can cover two octaves in bandwidth, can be appropriately manipulated in experimental arrangements and thus be used to synthesize extremely short pulses with durations in the single-cycle regime, which was also corroborated by our theoretical model.

A main characteristic of the generated beams is their broadband spectrum, which can result in pulses shorter than the pump beams. This characteristic has been deeply studied for the efficient generation and measurement of broadband pulses in the deep-ultraviolet region by highly nondegenerate four-wave mixing of fundamental and second-harmonic pulses from a Titanium:Sapphire laser amplifier. Our theoretical model also anticipates the generation of multiple broadband pulses extending into the vacuum ultraviolet by cascaded four-wave mixing in fluorides, using a novel multiple pump beam configuration. All these results help to establish the capability and potential of cascaded four-wave mixing for producing new sources consisting of a set of mutually coherent broadband pulses with different colors, whose characteristics are very difficult to obtain otherwise, based on simple nonlinear media, and which can have numerous applications in ultrafast spectroscopy and other fields of science and technology.

Acknowledgments: Helder M. Crespo gratefully acknowledges support from the Portuguese funding agency, Fundação para a Ciência e Tecnologia (FCT), through Project Grant PTDC/FIS/122511/2010 and Sabbatical Leave Grant SFRH/BSAB/105974/2015, co-funded by COMPETE and FEDER. Rosa Weigand acknowledges support from Grant GR3/14-910133 (Ayudas para los grupos de investigación UCM). The authors thank J. L. Silva for helpful comments.

Author Contributions: Helder M. Crespo performed pioneering work on this subject and both authors further developed it. The authors contributed equally to this paper.

Conflicts of Interest: The authors declare no conflict of interest.

References

1. Shen, Y.R. *The Principles of Nonlinear Optics*; John Wiley & Sons: Hoboken, NJ, USA, 2003.
2. Eichler, H.J.; Günter, P.; Pohl, D.W. Diffraction and Four-Wave Mixing Theory. In *Laser-Induced Dynamic Gratings*; Tamir, T., Ed.; Springer: Berlin, Germany, 1986; pp. 94–121.
3. Durfee, C.G.; Backus, S.; Kapteyn, H.C.; Murnane, M.M. Intense 8-fs pulse generation in the deep ultraviolet. *Opt. Lett.* **1999**, *24*, 697–699.
4. Penzkofer, A.; Lehmeir, H.J. Theoretical investigation of noncollinear phase-matched parametric four-photon amplification of ultrashort light pulses in isotropic media. *Opt. Quantum. Electron.* **1993**, *25*, 815–844.
5. Akmahnov, S.A.; Martynov, V.A.; Saltiel, S.M.; Tunkin, V.G. Observation of nonresonant six-photon processes in a calcite crystal. *JETP Lett.* **1975**, *22*, 65–67.
6. Lichtman, E.; Friesem, A.A.; Waarts, R.G.; Yaffe, H.H. Exact solution of four-wave mixing of copropagating light beams in a Kerr medium. *J. Opt. Soc. Am. B* **1987**, *4*, 1801–1805.
7. Hart, D.L.; Judy, A.; Brian Kennedy, T.A.; Roy, R.; Stoev, K. Conservation law for multiple four-wave-mixing processes in a nonlinear optical medium. *Phys. Rev. A* **1994**, *50*, 1807–1813.

8. Hart, D.L.; Judy, A.F.; Roy, R.; Beletic, J.W. Dynamical evolution of multiple four-wave-mixing processes in an optical fiber. *Phys. Rev. E* **1998**, *57*, 4757–4774.

9. Tan, H.; Banfi, G.P.; Tomaselli, A. Optical frequency mixing through cascaded second-order processes in β-barium borate. *Appl. Phys. Lett.* **1993**, *63*, 2472–2474.

10. Varanavicius, A.; Dubietis, A.; Berzanskis, A.; Danielius, R.; Piskarskas, A. Near-degenerate cascaded four-wave mixing in an optical parametric amplifier. *Opt. Lett.* **1997**, *22*, 1603–1605.

11. Crespo, H.; Mendonça, J.T.; Dos Santos, A. Cascaded highly nondegenerate four-wave-mixing phenomenon in transparent isotropic condensed media. *Opt. Lett.* **2000**, *25*, 829–831.

12. Chin, A.H.; Calderón, O.G.; Kono, J. Extreme Midinfrared Nonlinear Optics in Semiconductor. *Phys. Rev. Lett.* **2001**, *86*, 3292–3295.

13. Misoguti, L.; Backus, S.; Durfee, C.G.; Bartels, R.; Murnane, M.M.; Kapteyn, H.C. Generation of broadband VUV light using cascaded processes. *Phys. Rev. Lett.* **2001**, *87*, 013601.

14. Lu, C.; Yang, L.; Zhi, M.; Sokolov, A.; Yang, S.; Hsu, C.; Kung, A. Generation of octave-spanning supercontinuum by Raman-assisted four-wave mixing in single-crystal diamond. *Opt. Express* **2014**, *22*, 4075–4082.

15. Liu, J.; Kobayashi, T. Cascaded four-wave mixing and multicolored arrays generation in a sapphire plate by using two crossing beams of femtosecond laser. *Opt. Express* **2008**, *16*, 22119–22125.

16. Liu, W.; Zhu, L.; Wang, L.; Fang, C. Cascaded four-wave mixing for broadband tunable laser sideband generation. *Opt. Lett.* **2013**, *38*, 1772–1774.

17. Liu, J.; Kobayashi, T. Wavelength-tunable, multicolored femtosecond-laser pulse generation in fused-silica glass. *Opt. Lett.* **2009**, *34*, 1066–1068.

18. Zhang, X.; Giessen, H. Four-wave mixing based on cascaded second-order nonlinear processes in a femtosecond optical parametric oscillator operating near degeneracy. *Appl. Phys. B* **2004**, *79*, 441–447.

19. Li, Y.H.; Zhao, Y.Y.; Wang, L.J. Demonstration of almost octave-spanning cascaded four-wave mixing in optical microfibers. *Opt. Lett.* **2012**, *37*, 3441–3443.

20. Zhang, H.; Zhou, Z.; Lin, A.; Cheng, J.; Liu, H.; Si, J.; Chen, F.; Hou, X. Controllable cascaded four-wave mixing by two chirped femtosecond laser pulses. *Appl. Phys. B* **2012**, *108*, 487–491.

21. Weigand, R.; Mendonça, J.T.; Crespo, H.M. Cascaded nondegenerate four-wave-mixing technique for high-power single-cycle pulse synthesis in the visible and ultraviolet ranges. *Phys. Rev. A* **2009**, *79*, 063838.

22. Silva, J.L.; Weigand, R.; Crespo, H.M. Octave-spanning spectra and pulse synthesis by nondegenerate cascaded four-wave mixing. *Opt. Lett.* **2009**, *34*, 2489–2491.

23. Nisoli, M.; De Silvestri, S.; Svelto, O.; Szipöcs, R.; Ferencz, K.; Spielmann, Ch.; Sartania, S.; Krausz, F. Compression of high-energy laser pulses below 5 fs. *Opt. Lett.* **1997**, *22*, 522–524.

24. Harris, S.E.; Sokolov, A.V. Subfemtosecond Pulse Generation by Molecular Modulation. *Phys. Rev. Lett.* **1998**, *81*, 2894–2897.

25. Mendonça, J.T.; Crespo, H.; Guerreiro, A. A new method for high-harmonic generation by cascaded four-wave mixing. *Opt. Commun.* **2001**, *188*, 383–388.

26. Kobayashi, T.; Liu, J.; Okamura, K. Applications of parametric processes to high-quality multicolour ultrashort pulses, pulse cleaning and CEP stable sub-3fs pulse. *J. Phys. B* **2012**, *45*, 074005.

27. Silva, J.L.; Crespo, H.M.; Weigand, R. Generation of high-energy vacuum UV femtosecond pulses by multiple-beam cascaded four-wave mixing in a transparent solid. *Appl. Opt.* **2011**, *50*, 1968–1973.

28. Weigand, R.; Crespo, H.M. Generation of high-energy broadband femtosecond deep-ultraviolet pulses by highly nondegenerate noncollinear four-wave mixing in a thin transparent solid. *Appl. Phys. B* **2013**, *111*, 559–565.

29. Jailaubekov, E.A.; Bradforth, S.S. Tunable 30-femtosecond pulses across the deep ultraviolet. *Appl. Phys. Lett.* **2005**, *87*, 021107.

30. Fuji, T.; Horio, T.; Suzuki, T. Generation of 12fs deep-ultraviolet pulses by four-wave mixing through filamentation in neon gas. *Opt. Lett.* **2007**, *32*, 2481–2483.

31. Fuji, T.; Suzuki, T.; Serebryannikov, E.E.; Zheltikov, A. Experimental and theoretical investigation of a multicolor filament. *Phys. Rev. A* **2009**, *80*, 063822.

32. Kida, Y.J.; Liu, J.; Teramoto, T.; Kobayashi, T. Sub-10 fs deep-ultraviolet pulses generated by chirped-pulse four-wave mixing. *Opt. Lett.* **2010**, *35*, 1807–1809.

33. Backus, S.; Peatross, J.; Zeek, Z.; Rundquist, A.; Taft, G.; Murnane, M.M.; Kapteyn, H.C. 16-fs, 1-μJ ultraviolet pulses generated by third-harmonic conversion in air. *Opt. Lett.* **1996**, *21*, 665–667.

34. Graf, U.; Fiess, M.; Schultze, M.; Kienberger, R.; Krausz, F.; Goulielmakis, E. Intense few-cycle light pulses in the deep ultraviolet. *Opt. Express* **2008**, *16*, 18956–18963.

35. Reiter, F.; Graf, U.; Schultze, M.; Schweinberger, W.; Schröder, H.; Karpowicz, N.; Azzeer, A.; Kienberger, R.; Krausz, F.; Goulielmakis, E. Generation of sub-3 fs pulses in the deep ultraviolet. *Opt. Lett.* **2010**, *35*, 2248–2250.

36. Bertrand, J.B.; Wörner, H.J.; Bandulet, H.C.; Bisson, É.; Spanner, M.; Kieffer, J.C.; Villeneuve, D.M.; Corkum, P.B. Ultrahigh-Order Wave Mixing in Noncollinear High Harmonic Generation. *Phys. Rev. Lett.* **2011**, *106*, 023001.

Section 2:
Letters & Articles

Chapter 1:
Generation of Multi-Color Laser Emission

High-Energy, Multicolor Femtosecond Pulses from the Deep Ultraviolet to the Near Infrared Generated in a Hydrogen-Filled Gas Cell and Hollow Fiber

Kazuya Motoyoshi, Yuichiro Kida and Totaro Imasaka

Abstract: We investigate four-wave mixing in hydrogen gas using a gas cell and a hollow fiber for the generation of high-energy, multicolor femtosecond (fs) optical pulses. Both a hydrogen-filled gas cell and hollow fiber lead to the generation of multicolor fs pulses in a broad spectral range from the deep ultraviolet to the near infrared. However, there is a difference in the energy distribution of the multicolor emission between the gas cell and the hollow fiber. The hydrogen-filled gas cell generates visible pulses with higher energies than the pulses created by the hollow fiber. We have generated visible pulses with energies of several tens of microjoules. The hydrogen-filled hollow fiber, on the other hand, generates ultraviolet pulses with energies of a few microjoules, which are higher than the energies of the ultraviolet pulses generated in the gas cell. In both schemes, the spectral width of each emission line supports a transform-limited pulse duration shorter than 15 fs. Four-wave mixing in hydrogen gas therefore can be used for the development of a light source that emits sub-20 fs multicolor pulses in a wavelength region from the deep ultraviolet to the near infrared with microjoule pulse energies.

Reprinted from *Appl. Sci.* Cite as: Motoyoshi, K.; Kida, Y.; Imasaka, T. High-Energy, Multicolor Femtosecond Pulses from the Deep Ultraviolet to the Near Infrared Generated in a Hydrogen-Filled Gas Cell and Hollow Fiber. *Appl. Sci.* **2014**, 4, 318–330.

1. Introduction

Four-wave mixing (FWM) has been investigated and used in the past few decades to generate multicolor laser emission in various wavelength regions. In 1981, more than 40 laser emission lines spanning from the deep ultraviolet (DUV) to the near infrared (NIR) were generated by focusing a two-color nanosecond pulse into hydrogen gas [1]. The generation of multicolor laser emission via FWM has hitherto been extensively studied in deuterium and hydrogen gases [2–5] and extremely short optical pulses with durations shorter than 2 fs have been generated [3,5] by Fourier synthesis of the emission lines [6,7].

The generation of multicolor laser emission via FWM has also been investigated in the femtosecond (fs) regime [8–21]. Resonant and non-resonant FWM has been

73

investigated in bulk media [11,13,15–17], leading to multicolor laser emission with a pulse energy of about 1 microjoule [17]. By employing the pre-compression technique, multicolor emission with short pulse durations less than 20 fs has been demonstrated [16]. Such short multicolor pulses can be used in ultrafast spectroscopy [22–24], as well as in nonlinear optical microscopy [25]. The efficiency of the multicolor generation can be improved by more than one order of magnitude using gaseous media rather than bulk media. In addition, the spectral range from the DUV to the NIR has been covered by employing Raman active gases as the nonlinear media for FWM [9,10,12,14,20,21]. When pumping with short fs pulses with high intensities, self-phase modulation (SPM) and cross-phase modulation (XPM) are simultaneously induced, which results in a broad spectral width for each multicolor emission. The broad spectral width is advantageous for generating a short optical pulse after adequate dispersion compensation. The maximum spectral width of each multicolor emission is limited by the frequency separation between the center frequencies of the adjacent multicolor emission, which is determined by the Raman shift of the medium used. When using the vibrational transition of molecular hydrogen with a large Raman shift frequency of 4155 cm^{-1} [26], it is possible to generate multicolor fs pulses with spectral widths supporting transform-limited pulse durations of about 10 fs. Multicolor generation via the vibrational transition has been investigated both in a hydrogen-filled hollow fiber using two-color fs pump pulses at 800 and 600 nm [12] and in a hydrogen-filled gas cell using shorter two-color NIR pump pulses emitting at 800 and 1200 nm [20]. These investigations have reported multicolor laser emission covering from the DUV to NIR. The use of a hollow fiber [12] and gas cell [20], however, has been investigated under independent experimental conditions. Therefore, a quantitative discussion of the efficiency of multicolor generation and the characteristics of the pulses has been hindered. Information related to the pulse energy of each multicolor emission has not been reported to date; these data are important for evaluating the multicolor laser pulses.

We investigate here two approaches of FWM in a hydrogen-filled gas cell and a hydrogen-filled hollow fiber. We compare the results to analyze laser sources of multicolor fs pulses. By maintaining the experimental parameters at a constant—except for the gas pressure, which has been optimized in each experiment—the two schemes can be directly compared with one other. We find that the gas-filled hollow fiber with a longer propagation distance than that of the gas cell does not always lead to higher efficiencies in the multicolor generation. The gas cell gives rise to visible multicolor pulses with higher pulse energies than those generated in the hollow fiber. On the other hand, the latter produces higher-energy multicolor pulses in the ultraviolet range.

2. Experimental Section

Figure 1 shows the experimental setup. A part of the NIR pulse emerging from a Ti:sapphire regenerative amplifier (800 nm, 35 fs, 4 mJ, 1 kHz, Legend Elite-USP, Coherent Inc., Santa Clara, CA, USA) was used as the pump source for an optical parametric amplifier (OPA, OPerASolo, Coherent Inc., Santa Clara, CA, USA). The OPA generated a NIR pulse at 1200 nm, whose frequency separation from the Ti:sapphire amplifier (800 nm) was adjusted to the vibrational Raman shift frequency of molecular hydrogen (4155 cm^{-1}). The output of the OPA (pump 1; P1) was spatially combined with the remaining part of the output beam of the regenerative amplifier at 800 nm (pump 2; P2). The beam was focused with an off-axis parabolic mirror (with a focal length of 655 mm) into a gas cell filled with hydrogen or a fused-silica hollow fiber (core diameter, 320 µm; length, 600 mm) placed inside a gas cell (length, 1 m) filled with hydrogen (hereafter denoted as a hollow fiber chamber). The core diameter of the hollow fiber of 320 µm was chosen for keeping the same focusing condition as for the gas cell, and the fiber length of 600 mm was the maximum possible length for the hollow fiber chamber. The gas cell and the hollow fiber chamber were equipped with 0.5-mm-thick windows made of fused silica. The time delay between the two pulses (P1 and P2) was optimized to obtain the highest energy in the highest-order anti-Stokes Raman emission generated at a gas pressure of 0.4 atm. The spectra of the output beams from the gas cell and hollow fiber chamber were measured using a multichannel spectrometer (Maya2000pro, Ocean Optics, Dunedin, FL, USA). The spectral response was calibrated using a deuterium-halogen light source over a wavelength range of 220–900 nm.

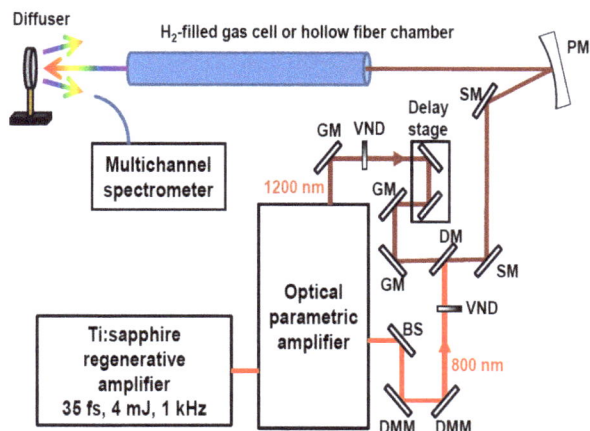

Figure 1. The experimental setup. GM, gold mirror; SM, silver mirror; DM, dichroic mirror; PM, off-axis parabolic mirror; BS, beam splitter; DMM, dielectric multi-layer mirror; VND, variable neutral density filter.

3. Results and Discussion

3.1. Generation of the Multicolor Emission in the Gas Cell and the Gas-Filled Hollow Fiber

When we used a hydrogen-filled gas cell, the FWM efficiently occurred in the vicinity of the foci at which the input beams have high intensities. The confocal parameter $(2\pi W_0{}^2/\lambda)$ is a measure of the interaction length, where W_0 and λ are the beam waist radius and the wavelength of the laser beam, respectively. From the measured focal beam diameters of P1 and P2—300 μm and 200 μm—the confocal parameters of the beams were calculated. The value for P2 was 80 mm, which was shorter than that of P1 (120 mm). The interaction length for the gas cell was, therefore, estimated to be approximately 80 mm, which was shorter than that of the hydrogen-filled hollow fiber.

The energies of the input and output pulses measured in front of the evacuated gas cell and hollow fiber chamber are listed in Table 1. From these values, the transmittance of the gas cell was calculated to be 90% for both P1 and P2. This value is in good agreement with the value calculated from Fresnel losses on the surfaces of the fused-silica windows of the gas cell. On the other hand, the throughput for the hollow fiber chamber was calculated to be 50% and 67% for P1 and P2, respectively. The smaller throughput for the hollow fiber may be due to coupling losses at the entrance of the hollow fiber and the linear propagation losses inside the fiber. For propagation of the EH_{11} mode inside the hollow fiber, the linear propagation loss through the hollow fiber is calculated to be 9% at 1200 nm and 4% at 800 nm [27]. By assuming a coupling into the EH_{11} mode and propagation of the mode inside the hollow fiber, the efficiency of coupling at the entrance of the hollow fiber is calculated to be 61% and 78% for P1 and P2, respectively. The smaller coupling efficiency for P1 than P2 arises from the too large beam diameter of P1 (300 μm) to be perfectly coupled into the hollow fiber.

Table 1. Input and output energies of P1 and P2.

	Gas Cell		Hollow Fiber	
	Input	Output	Input	Output
P1, 1200 nm	262 μJ	236 μJ	278 μJ	140 μJ
P2, 800 nm	245 μJ	220 μJ	252 μJ	168 μJ

The pulse durations of the input pulses were measured based on cross-correlation frequency-resolved optical gating, which allows for the characterization of the two unknown pulses simultaneously [28]. The pulse durations obtained were 50–60 fs.

The spectra of the output beams measured at different pressures for both the gas cell and the hollow fiber chamber are shown in Figure 2. The efficiency in the generation of the anti-Stokes emission increased with increasing gas pressure for both cases. The spectral width of the emission increased because of SPM and XPM induced by the intense input pump pulses [9,10,12,14]. In the case of the gas cell, the anti-Stokes emission lines were well separated from each other in the spectrum at pressures ranging up to 2 atm, while the spectrum became continuous above 1 atm in the case of the hollow fiber. The difference could be explained by the longer interaction length in the hollow fiber than in the gas cell. The spectrum of the output beam from the hollow fiber became a continuum at 2 atm, as shown in Figure 3. Even at a pressure of 0.8 atm, the influence of the phase modulation was appreciable in the case of the hollow fiber. An image of the output beam from the hollow fiber is shown in Figure 4, which was taken after passing the beam through a fused-silica prism and projecting the separated beam on a white screen. The spots are not separated from each other because of the broad bandwidth of the emission induced by the phase modulations. The spectral width of the anti-Stokes emission generated in the hollow fiber supported a sub-10 fs pulse duration, even at a pressure of 1 atm; no spectral overlap between adjacent anti-Stokes emission features was observed. Similar results were reported by Sali *et al.*, in which these authors used longer pump pulses emitting at shorter wavelengths than in this work [12]. The hollow fiber thus may be able to generate sub-10 fs multicolor laser pulses covering a wavelength region from the DUV to NIR by relying on proper dispersion compensation for each emission feature.

Although the efficiency of the XPM is smaller and the spectral width of anti-Stokes emission is narrower for the gas cell, the spectral widths of the multicolor emission achieved for the gas cell at a pressure of 2 atm support transform-limited pulse durations shorter than 15 fs. The spectral widths of the emission were 1200 cm^{-1} (AS1), 1300 cm^{-1} (AS2), 1200 cm^{-1} (AS3), and 1000 cm^{-1} (AS4). Similar to the case of the hollow fiber, the generation of sub-20 fs multicolor pulses may be possible after appropriate dispersion compensation.

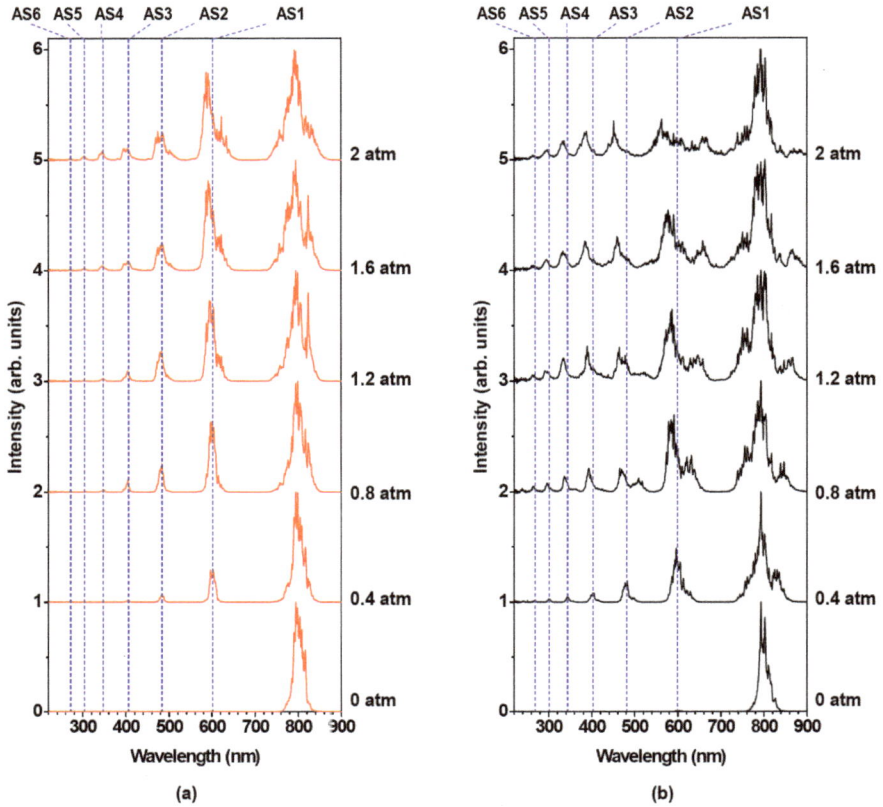

Figure 2. The spectra of output beams at different hydrogen pressures. (**a**) Gas cell; (**b**) Hollow fiber.

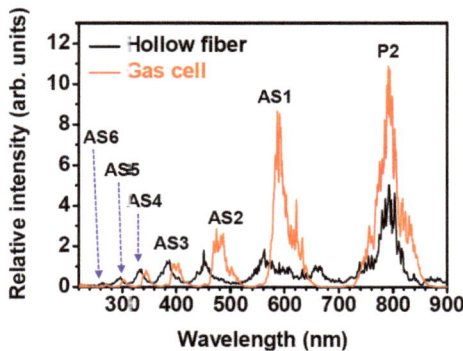

Figure 3. Spectra of output beams from the hollow fiber (black solid line) and the gas cell (red solid line) at a pressure of 2 atm. The spectra for P1 are not shown in the figure.

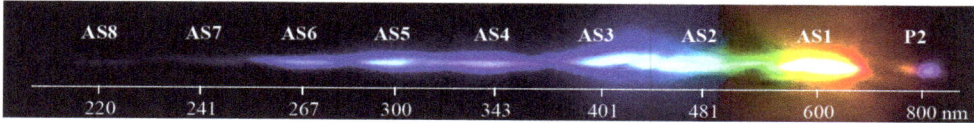

Figure 4. Photograph of the multicolor emission.

3.2. Energies of the Multicolor Emission

For estimation of the energy of the anti-Stokes emission, the gas cell was first evacuated. Only P2 was focused into the gas cell and the pulse energy of P2 emerging from the cell was measured using a power meter. The spectrum of the output P2 was measured using a sensitivity-calibrated spectrometer, the position of which was kept unchanged during the subsequent experimental procedure. The measured pulse energy was divided by the integral of the spectral intensity of P2. The resulting value is hereafter referred to as the calibration factor. The gas cell was then filled with hydrogen gas and both P1 and P2 were focused into the gas cell. The spectrum of the output beam from the cell was measured using the spectrometer at different gas pressures. The integral of the spectral intensity for each anti-Stokes emission was multiplied by the calibration factor to estimate the pulse energy of the emission. The same procedure was repeated for the hydrogen-filled hollow fiber. Because of the nearly continuous structure of the spectrum observed at pressures higher than 1 atm, the pulse energy could not be estimated for the hydrogen-filled hollow fiber at high pressures.

The results of the energy estimation are shown in Figure 5. In both the experiments using the gas cell and the hollow fiber, the energy of P2 was notably depleted by the generation of the anti-Stokes emission. The energies of the low-order anti-Stokes emission, *i.e.*, AS1 and AS2, generated in the gas cell were 1.8 times higher than those generated in the hollow fiber. This result indicates that the gas cell is superior to the gas-filled hollow fiber in generating high-energy visible pulses. The conversion efficiency from P2 into the anti-Stokes emission was similar in these two cases (see the observed conversion efficiencies shown in Figure 5). This fact suggests that if there are no propagation and coupling losses for the hollow fiber, the anti-Stokes emission of AS1 and AS2 would have energies close to those for the gas cell. The energies of the higher-order anti-Stokes emission, on the other hand, were larger for the hollow fiber than those for the gas cell. The conversion efficiencies from P2 to the anti-Stokes emission of AS5 and AS6 were three times and four times larger than those generated in the gas cell, respectively, as shown in Figure 5c,f.

The energies of P1 and P2 for the gas cell were reduced until the output energies were the same as those in the hollow fiber experiment (140 µJ for P1 and 160 µJ for P2). For the energy reduction, the neutral density filters placed in the input beam paths were employed (see Figure 1). The filters with thicknesses of 2 mm would not have stretched the pump pulses appreciably; each filter stretches a 35-fs pulse at 800 nm by only 0.5 fs. The resultant spectrum of the output beam from the gas cell pressurized at 2 atm is shown in Figure 6, together with the spectrum measured using the hollow fiber without a reduction in the pump energy. No notable difference was observed in the energies of the first- and second-order anti-Stokes emission under these conditions, although the spectral intensities of the emission were higher for the gas cell. This result suggests that the pulse energy of the visible emission from the hollow fiber does not exceed that from the gas cell, even under the ideal condition that there are no propagation and coupling losses for the hollow fiber.

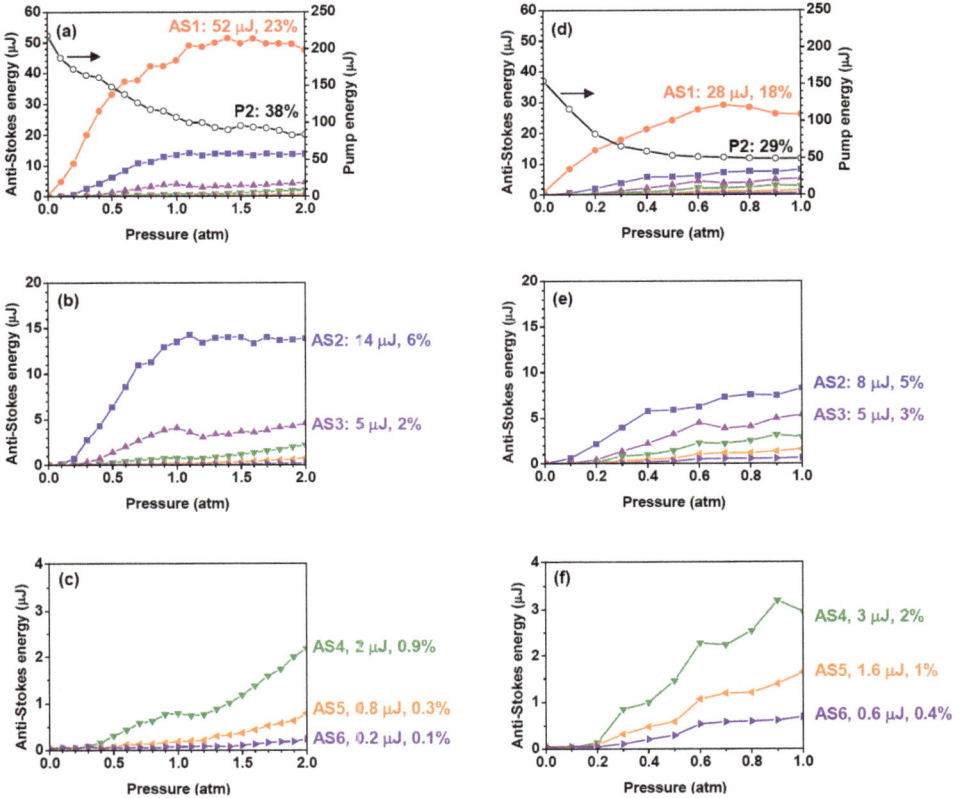

Figure 5. The pressure dependence of the energy for the output beams. (a–c) The gas cell and (d–f) the hollow fiber. The maximum energy obtained for each anti-Stokes emission is indicated together with the conversion efficiency from P2 to the anti-Stokes emission. For P2, the ratio of the output energy measured at a specified pressure and the output energy measured at 0 atm is listed in each panel.

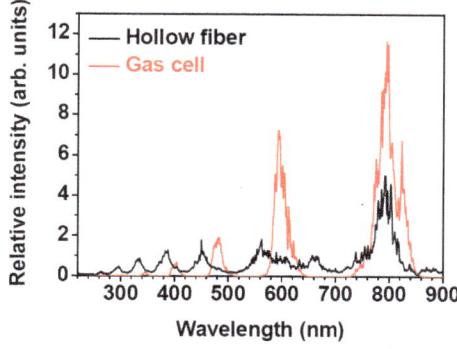

Figure 6. Spectra of the output beam from the gas cell (red solid line) and the hollow fiber (black solid line) measured at a pressure of 2 atm. The energies of the input pump pulses for the gas cell are lower than those for the hollow fiber.

3.3. Four-Wave Mixing

The efficiency in FWM is determined by parameters such as the intensities of the pump pulses, phase mismatch, and the interaction length. Under the assumption that no depletion occurs for pump pulses, the intensity of the first-order anti-Stokes emission, I_{AS1}, can be described as [29,30]:

$$I_{AS1} = \frac{\epsilon_0 n \omega_{AS1}{}^2 n_2{}^2 z^2}{2c} |\varepsilon_{P1}|^2 |\varepsilon_{P2}|^4 \mathrm{sinc}^2(-\frac{\Delta\beta z}{2}) \qquad (1)$$

where ε_0, c, n, n_2, z, and ω_{AS1}, are the vacuum permittivity, the velocity of light in vacuum, the linear and nonlinear refractive indices of the gas, the interaction length, and the angular frequency of AS1, respectively. The intensities of P1 (I_{P1}) and P2 (I_{P2}) are proportional to the square of the corresponding electric field amplitudes, ε_{P1} and ε_{P2}, respectively. The phase mismatch $\Delta\beta$ in the equation is expressed as $\beta_{P1} + \beta_{AS1} - 2\beta_{P2}$, where β_{P1}, β_{AS1}, and β_{P2} stand for the propagation constants for P1, AS1, and P2. The phase mismatch in the hollow fiber can be calculated by taking into account the contributions of the gas and the waveguide [27,29,30].

In the generation of multicolor beams through cascaded FWM, the anti-Stokes emission is generated in the first step and higher-order anti-Stokes emission is then generated during the propagation in the gas. The energy of the first anti-Stokes emission is higher than that of the high-order anti-Stokes emission and it is a good measure of the efficiency of the multicolor generation.

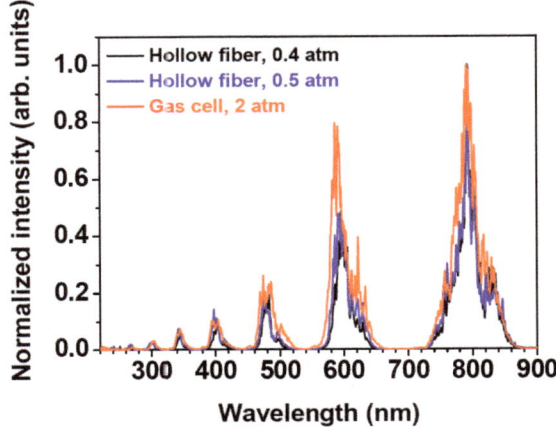

Figure 7. Spectra of the output beam from the hollow fiber at 0.4 atm (black solid line) and 0.5 atm (blue solid line) and the gas cell (red solid line) at 2 atm. The spectrum is normalized by the intensity of the highest peak.

Equation (1) can be used to calculate the gas pressure in the hollow fiber, which leads to a similar efficiency of multicolor generation as that in the gas cell with a pressure of 2 atm. The value of $\Delta\beta z/2$ in Equation (1) is calculated to be 0.66 rad for the gas cell and hence the value of $\mathrm{sinc}^2(-\Delta\beta z/2)$ can be approximated as unity. From Equation (1) and the fact that the nonlinear refractive index is proportional to the gas pressure, the conversion efficiency from the pump to the first-order anti-Stokes emission, I_{AS1}/I_{P2}, is proportional to $I_{P1}I_{P2}p^2z^2$, where p is the gas pressure. The output energies of P1 and P2 in the case of the hollow fiber were 0.59 and 0.76 times smaller than those for the gas cell, respectively, as indicated in Table 1. The value of $I_{P1}I_{P2}$ for the hollow fiber is hence considered to be 0.43 times smaller than that for the gas cell. From this fact, together with the ratio of the interaction lengths in the hollow fiber and the gas cell (600 mm/80 mm), the gas pressure in the hollow fiber, which provides the same value of I_{AS1}/I_{P2} as for the gas cell, is calculated to be 0.4 atm. In Figure 7, we show the spectra of the output beams from the hollow fiber at 0.4 and 0.5 atm and the gas cell at 2 atm for comparison. Excellent agreement is observed between the spectra for the hollow fiber at 0.4–0.5 atm and for the gas cell at 2 atm, although no pump depletion is assumed to occur in the model used in Equation (1).

3.4. Phase Mismatch in the Generation of High-Order Anti-Stokes Emission

In Table 2, the values of $\Delta\beta z/2$ in the generation of anti-Stokes emission are shown for the pathways with the energy conservations of $-\omega_{P1} + \omega_{P2} + \omega_{ASN-1} - \omega_{ASN} = 0$, where ω_{P1}, ω_{P2}, ω_{ASN-1}, and ω_{ASN} are the angular frequencies of P1, P2,

$(N - 1)^{\text{th}}$-order anti-Stokes, and N^{th}-order anti-Stokes emission. In the calculation, the propagation length is assumed to be 600 mm for the hollow fiber and 80 mm for the gas cell and the gas pressures are assumed to be 1 atm and 2 atm, respectively. The value of $\Delta\beta z/2$ for AS1 is lower than π in the both cases of the gas cell and hollow fiber. As shown in Figure 5, the conversion efficiency from P2 to AS1 saturated at pressures higher than 1 atm and 0.5 atm for the gas cell and the hollow fiber, respectively. The saturated conversion efficiency from P2 to AS1 was similar in both cases (23% for the gas cell and 18% for the hollow fiber). From these facts, the saturation would be related to the consumption of the energy for the generation of high-order anti-Stokes emission rather than the phase mismatch. In other words, the waveguide dispersion of the hollow fiber does not have notable influence to the conversion efficiency from P2 to AS1. Change in the parameters of the hollow fiber, such as use of a longer hollow fiber and a hollow fiber with a smaller core diameter, therefore, would not lead to a higher conversion efficiency to AS1. It just shifts the saturation pressure to a lower pressure. The energy of the visible multi-color emission (AS1) emitted from a gas cell is therefore expected to be always higher than that from a hollow fiber.

Table 2. Calculated value of $\Delta\beta z/2$ for anti-Stokes emission.

| | $\Delta\beta z/2$ (rad) | | | | | |
	AS1	AS2	AS3	AS4	AS5	AS6
Hollow fiber, 1 atm	1.39	4.36	8.61	14.14	21.15	29.89
Gas cell, 2 atm	0.66	1.57	2.75	4.26	6.15	8.49

The values of the $\Delta\beta z/2$ are non-negligible for all the anti-Stokes emission and exceed π, except AS1 for the hollow fiber and except AS1, AS2, and AS3 for the gas cell. This fact does not explain the increase in the energy of the high-order anti-Stokes emission with increasing gas pressure, which was observed in the experiment (Figure 5c). For high-order anti-Stokes emission, other pathways therefore need to be considered.

There are several possible pathways useful for the generation of the high-order anti-Stokes emission. For instance, four paths——$\omega_{P1} + \omega_{P2} + \omega_{AS3} - \omega_{AS4} = 0$, $-\omega_{P2} + \omega_{AS1} + \omega_{AS3} - \omega_{AS4} = 0$, $-\omega_{AS1} + \omega_{AS2} + \omega_{AS3} - \omega_{AS4} = 0$, and $-\omega_{AS2} + \omega_{AS3} + \omega_{AS3} - \omega_{AS4} = 0$——would contribute to the generation of the fourth-order anti-Stokes emission, AS4. Among them, the phase mismatch, $\Delta\beta$, is the highest for the first pathway and lowest for the fourth pathway. $\Delta\beta z/2$ is calculated to be 1.5 rad for the fourth path in the case of the gas cell with a pressure of 2 atm. This value is lower than $\pi/2$ and the path would contribute to the increase in the anti-Stokes signal intensity at high pressures. Since the degree of phase mismatch is proportional to the gas pressure, the path most effective for the generation of the Raman emission would change depending on the gas pressure, even under the same experimental conditions for the propagation length and the input pulse intensities. This effect may be of importance, particularly in the DUV, at which numerous possible pathways are plausible for the generation of Raman emission. For the gas-filled hollow fiber, it should be necessary to take into account the contributions from high-order propagation modes. For several combinations of high-order propagation modes, the gas pressure at which phase matching is satisfied becomes higher than the phase-matching gas pressure for the lowest-order propagation modes [29–32].

3.5. Spectral Blueshift in the Ultraviolet Sidebands

A spectral blueshift was observed for the anti-Stokes emission in the ultraviolet region when a hollow fiber was filled at pressures higher than 0.4 atm. As can be seen in Figure 5b, the wavelengths of the emission are continuously blueshifted with increasing gas pressure. This blueshift was not observed for the gas cell at pressures up to 2 atm. The blueshift can be explained in terms of the pulse chirp of each emission. In Figure 8, we calculate the group delay of each emission with respect to P2 for (a) the gas cell and (b) the hollow fiber under the propagation mode of EH_{11} after the propagation length of $z = 80$ and 600 mm, respectively. The results suggest that the group delay of the high-order anti-Stokes emission in the hollow fiber (AS4–6) significantly increases with an increase in gas pressure. The anti-Stokes emission delayed with respect to the pump pulses would temporally overlap at the trailing edges of the pump pulses and the phase could be modulated via XPM induced by the pump pulses, resulting in a spectral blueshift. On the other hand, the group delay is small in the gas cell for all orders of the anti-Stokes emission. Thus, the group delay could have a negligible effect on the frequency of the anti-Stokes emission in the case of the gas cell, as shown in the experimental results of Figure 2a.

The group delays of AS4, AS5, and AS6 exceed the pulse durations of the pump pulses of 50 fs at pressures higher than 1.2 atm in the case of the hollow fiber (Figure 8b). These delays make the interaction between pump pulses and the high-order anti-Stokes pulses difficult for the generation of higher-order anti-Stokes emission via FWM. For this reason, combinations of frequency components with small phase mismatches for FWM should be effective for the generation of high-order anti-Stokes emission at high pressures for the hollow fiber (Figure 5f).

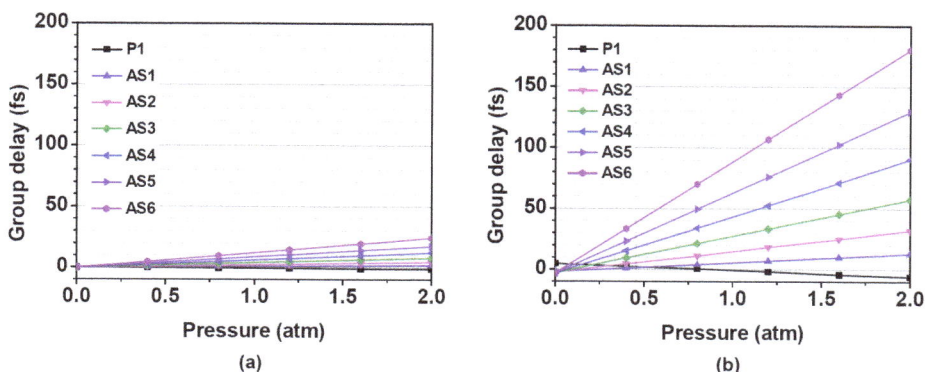

Figure 8. Group delay calculated for each emission against that of P2 for the gas cell (**a**) and the hollow fiber (**b**) at different gas pressures.

4. Conclusions

We have compared two approaches using a gas cell and a hollow fiber to generate high-energy multicolor fs pulses. The gas cell provides multicolor emission in the visible range with higher energies than those obtained using the hollow fiber. The pulse energies were several tens of microjoules, which were one order of magnitude larger than those obtained based on FWM in bulk media. Because of the simultaneous generation of SPM and XPM, the spectral bandwidth of the anti-Stokes emission increased with increasing gas pressure. The resultant bandwidth using the gas cell supports transform-limited pulse durations shorter than 15 fs. A broader bandwidth was obtained for multicolor emission generated in the gas-filled hollow fiber. In this case, multicolor laser pulses obtained in the DUV region had higher energies than pulses obtained using the gas cell, where the energy range was a few microjoules. The high-energy visible pulses generated in the gas cell are suitable for spectroscopic studies in the gas phase, such as multiphoton ionization of aromatic compounds and condensed-phase compounds. The ultraviolet pulses generated by the gas-filled hollow fiber, on the other hand, may be applied to spectroscopy of liquid and solid phases, such as the ultrafast transient absorption spectroscopy in biologically significant molecules [33].

Acknowledgments: This research was supported by the Japan Society for the Promotion of Science (JSPS) KAKENHI Grant No. 23245017.

Author Contributions: Drafting of manuscript: Kazuya Motoyoshi, Yuichiro Kida; Acquisition of data: Kazuya Motoyoshi; Analysis and interpretation of data: Kazuya Motoyoshi, Yuichiro Kida; Critical revision: Yuichiro Kida, Totaro Imasaka; Planning and supervision of the research: Yuichiro Kida, Totaro Imasaka.

References

1. Imasaka, T.; Kawasaki, S.; Ishibashi, N. Generation of More than 40 Laser Emission Lines from the Ultraviolet to the Visible Regions by Two-Color Stimulated Raman Effect. *Appl. Phys. B* **1989**, *49*, 389–392.

2. Yavuz, D.; Walker, D.; Shverdin, M.; Yin, G.; Harris, S. Quasiperiodic Raman Technique for Ultrashort Pulse Generation. *Phys. Rev. Lett.* **2003**, *91*, 233602.

3. Shverdin, M.; Walker, D.; Yavuz, D.; Yin, G.; Harris, S. Generation of a Single-Cycle Optical Pulse. *Phys. Rev. Lett.* **2005**, *94*, 033904.

4. Burzo, A.M.; Chugreev, A.V.; Sokolov, A.V. Optimized Control of Generation of Few Cycle Pulses by Molecular Modulation. *Opt. Commun.* **2006**, *264*, 454–462.

5. Chen, W.-J.; Hsieh, Z.-M.; Huang, S.; Su, H.-Y.; Lai, C.-J.; Tang, T.-T.; Lin, C.-H.; Lee, C.-K.; Pan, R.-P.; Pan, C.-L.; *et al.* Sub-Single-Cycle Optical Pulse Train with Constant Carrier Envelope Phase. *Phys. Rev. Lett.* **2008**, *100*, 5–8.

6. Yoshikawa, S.; Imasaka, T. A New Approach for the Generation of Ultrashort Optical Pulses. *Opt. Commun.* **1993**, *96*, 94–98.

7. Kaplan, A. Subfemtosecond Pulses in Mode-Locked 2π Solitons of the Cascade Stimulated Raman Scattering. *Phys. Rev. Lett.* **1994**, *73*, 1243–1246.

8. Irie, Y.; Imasaka, T. Generation of Vibrational and Rotational Emissions by Four-Wave Raman Mixing Using an Ultraviolet Femtosecond Pump Beam. *Opt. Lett.* **1995**, *20*, 2072–2074.

9. Kawano, H.; Hirakawa, Y.; Imasaka, T. Generation of More than 40 Rotational Raman Lines by Picosecond and Femtosecond Ti:sapphire Laser for Fourier Synthesis. *Appl. Phys. B* **1997**, *65*, 1–4.

10. Kawano, H.; Hirakawa, Y.; Imasaka, T. Generation of High-Order Rotational Lines in Hydrogen by Four-Wave Raman Mixing in the Femtosecond Regime. *IEEE J. Quantum Electron.* **1998**, *34*, 260–268.

11. Crespo, H.; Mendonça, J.T.; Dos Santos, A. Cascaded Highly Nondegenerate Four-Wave-Mixing Phenomenon in Transparent Isotropic Condensed Media. *Opt. Lett.* **2000**, *25*, 829–831.

12. Sali, E.; Kinsler, P.; New, G.; Mendham, K.; Halfmann, T.; Tisch, J.; Marangos, J. Behavior of High-Order Stimulated Raman Scattering in a Highly Transient Regime. *Phys. Rev. A* **2005**, *72*, 013813.

13. Zhi, M.; Sokolov, A.V. Broadband Coherent Light Generation in a Raman-Active Crystal Driven by Two-Color Femtosecond Laser Pulses. *Opt. Lett.* **2007**, *32*, 2251–2253.

14. Turner, F.C.; Trottier, A.; Strickland, D.; Losev, L.L. Transient Multi-Frequency Raman Generation in SF_6. *Opt. Commun.* **2007**, *270*, 419–423.

15. Liu, J.; Zhang, J.; Kobayashi, T. Broadband Coherent Anti-Stokes Raman Scattering Light Generation in BBO Crystal by Using Two Crossing Femtosecond Laser Pulses. *Opt. Lett.* **2008**, *33*, 1494–1496.

16. Liu, J.; Kobayashi, T. Generation of Sub-20-fs Multicolor Laser Pulses Using Cascaded Four-Wave Mixing with Chirped Incident Pulses. *Opt. Lett.* **2009**, *34*, 2402–2404.

17. Liu, J.; Kobayashi, T. Generation of μJ Multicolor Femtosecond Laser Pulses Using Cascaded Four-Wave Mixing. *Opt. Express* **2009**, *17*, 4984–4990.

18. Weigand, R.; Mendonça, J.; Crespo, H. Cascaded Nondegenerate Four-Wave-Mixing Technique for High-Power Single-Cycle Pulse Synthesis in the Visible and Ultraviolet Ranges. *Phys. Rev. A* **2009**, *79*, 063838.

19. Silva, J.; Crespo, H.; Weigand, R. Generation of High-Energy Vacuum UV Femtosecond Pulses by Multiple-Beam Cascaded Four-Wave Mixing in a Transparent Solid. *Appl. Opt.* **2011**, *50*, 1968–1973.

20. Shitamichi, O.; Imasaka, T. High-Order Raman Sidebands Generated from the near-Infrared to Ultraviolet Region by Four-Wave Raman Mixing of Hydrogen Using an Ultrashort Two-Color Pump Beam. *Opt. Express* **2012**, *20*, 27959–27965.

21. Cui, Z.; Chaturvedi, M.; Tian, B.; Ackert, J.; Turner, F.C.; Strickland, D. Spectral Red-Shifting of Multi-Frequency Raman Orders. *Opt. Commun.* **2013**, *288*, 118–121.

22. Kobayashi, T.; Saito, T.; Ohtani, H. Real-Time Spectroscopy of Transition States in Bacteriorhodopsin during Retinal Isomerization. *Nature* **2001**, *414*, 531–534.

23. Lanzani, G.; Cerullo, G.; Brabec, C.; Sariciftci, N. Time Domain Investigation of the Intrachain Vibrational Dynamics of a Prototypical Light-Emitting Conjugated Polymer. *Phys. Rev. Lett.* **2003**, *90*, 047402.

24. Yabushita, A.; Kobayashi, T. Primary Conformation Change in Bacteriorhodopsin on Photoexcitation. *Biophys. J.* **2009**, *96*, 1447–1461.

25. Rankin, B.R.; Kellner, R.R.; Hell, S.W. Stimulated-Emission-Depletion Microscopy with a Multicolor Stimulated-Raman-Scattering Light Source. *Opt. Lett.* **2008**, *33*, 2491–2493.

26. Stoicheff, B.P. High Resolution Raman Spectroscopy of Gases: IX. Spectra of H_2, HD, and D_2. *Can. J. Phys.* **1957**, *35*, 730–741.

27. Marcatili, E.A.J.; Schmeltzer, R.A. Hollow Metallic and Dielectric Waveguides for Long Distance Optical Transmission and Lasers. *Bell Syst. Tech. J.* **1964**, *43*, 1783–1809.

28. Kida, Y.; Nakano, Y.; Motoyoshi, K.; Imasaka, T. Frequency-Resolved Optical Gating with Two Nonlinear Optical Processes. *Opt. Lett.* **2014**, *39*, 3006–3009.

29. Durfee, C.G.; Misoguti, L.; Backus, S.; Kapteyn, H.C.; Murnane, M.M. Phase Matching in Cascaded Third-Order Processes. *J. Opt. Soc. Am. B* **2002**, *19*, 822–831.

30. Kida, Y.; Imasaka, T. Optical Parametric Amplification of a Supercontinuum in a Gas. *Appl. Phys. B* **2013**.

31. Misoguti, L.; Backus, S.; Durfee, C.; Bartels, R.; Murnane, M.; Kapteyn, H. Generation of Broadband VUV Light Using Third-Order Cascaded Processes. *Phys. Rev. Lett.* **2001**, *87*, 013601.

32. Faccio, D.; Grün, A.; Bates, P.K.; Chalus, O.; Biegert, J. Optical Amplification in the near-Infrared in Gas-Filled Hollow-Core Fibers. *Opt. Lett.* **2009**, *34*, 2918–2920.

33. Kobayashi, T.; Kida, Y. Ultrafast Spectroscopy with Sub-10 fs Deep-Ultraviolet Pulses. *Phys. Chem. Chem. Phys.* **2012**, *14*, 6200–6210.

Effect of Two-Photon Stark Shift on the Multi-Frequency Raman Spectra

Hao Yan and Donna Strickland

Abstract: High order Raman generation has received considerable attention as a possible method for generating ultrashort pulses. A large number of Raman orders can be generated when the Raman-active medium is pumped by two laser pulses that have a frequency separation equal to the Raman transition frequency. High order Raman generation has been studied in the different temporal regimes, namely: adiabatic, where the pump pulses are much longer than the coherence time of the transition; transient, where the pulse duration is comparable to the coherence time; and impulsive, where the bandwidth of the ultrashort pulse is wider than the transition frequency. To date, almost all of the work has been concerned with generating as broad a spectrum as possible, but we are interested in studying the spectra of the individual orders when pumped in the transient regime. We concentrate on looking at extra peaks that are generated when the Raman medium is pumped with linearly chirped pulses. The extra peaks are generated on the low frequency side of the Raman orders. We discuss how linear Raman scattering from two-photon dressed states can lead to the generation of these extra peaks.

Reprinted from *Appl. Sci.* Cite as: Yan, H.; Strickland, D. Effect of Two-Photon Stark Shift on the Multi-Frequency Raman Spectra. *Appl. Sci.* **2014**, *4*, 390–401.

1. Introduction

The nonlinear process of multi-frequency Raman generation (MRG) holds the promise of generating high intensity pulses with duration of just single femtoseconds. Such pulses could be used in nonlinear optical experiments for the study of molecular dynamics. MRG generates an ultrabroad spectrum containing several discrete frequency peaks by strongly driving a Raman transition of a molecule using two pump pulses. The peak frequencies of the two pumps are tuned such that the frequency difference equals the Raman frequency. This process was first observed by Imasaka and co-workers, when they generated several vibrational and rotational Raman orders in hydrogen, when their pump laser was tuned such that two frequency components were output [1].

The advantage of MRG over continuum generation for short pulse generation is that the phase only has to be corrected over several orders rather than the entire spectrum. On the other hand, because the spectrum is discrete, it generates a train of pulses. A pulse train of 1.6 fs pulses was obtained by simply phasing 7 Raman orders from hydrogen using a prism dispersion line and a liquid crystal phase modulator [2]. An even simpler compression technique of placing a few dispersive elements in the path has been shown to create trains of sub-fs pulses [3]. The duration of the pulse train is given by the pulse duration of the pump pulses and so shorter pump pulses would result in fewer pulses in the pulse train leading to more intense pulses. The 1.6 fs pulse train was pumped in the adiabatic regime with nanosecond long pulses. It has been shown that the MRG process can be efficient in the transient regime, where the pump pulses have durations of the coherence time of the Raman transition [4]. However, with short pulses, the MRG process competes with self-phase modulation (SPM), which creates a continuum spectrum under the discrete Raman spectra [5,6]. To avoid SPM, the short pulses are lengthened by linearly frequency chirping the pulses. We have previously observed that, when using linearly chirped pump pulses, extra peaks appear in the MRG spectrum of sulfur hexafluouride (SF_6) [7,8]. These peaks appear on the lower frequency side of the Raman peaks. These red shifted peaks also appear in the MRG spectra with chirped pulse pumping for the gas Raman media of hydrogen [5] as well as in the solid Raman medium, lead tungstate [9], indicating that the origin of the peaks is a fundamental process rather than being material dependent.

In order to determine if these red-shifted shoulders were simply the result of four-wave mixing, we have measured the MRG spectra as a function of tuning the instantaneous frequency separation [8]. We showed that the red shifted peaks shift further to the red as the instantaneous frequency separation of the two pump beams is reduced below the Raman frequency until the red-shift saturates. However, the peaks do not blue shift as the frequency separation increases above the Raman frequency, indicating that the shift is not due simply to non-resonant four-wave mixing. If the shifted peaks are not due to non-resonant four-wave mixing then it is most likely due to a Stark shift of the levels. Sokolov and co-workers noted that when using long pulses in the adiabatic regime, many more Raman orders were observed when the frequencies were tuned to be slightly closer together than the Raman resonance frequency [10]. When the frequencies were further apart than the Raman transition frequency, there was little difference in the generated spectra compared to on-resonance pumping. The authors of this work noted that the enhancement was given for in the in-phase coherent state, but not the anti-phased coherent state.

Hickman and co-authors derived the two-photon driven coherent state for Raman generation [11]. They derived the two-photon optical Bloch equations and showed that the states are split into two coherent states separated by the two-photon Rabi frequency Ω', which is given by:

$$\Omega'^2 = \Omega^2 + \Delta^2 \tag{1}$$

where:

$$\Omega e^{i\theta} = \frac{\alpha_{12}}{2\hbar} \sum_j V_j V_{j-1}^* \qquad (2)$$

$$I = \frac{c}{8\pi} \sum_j V_j V_{j-1} \qquad (3)$$

$$\Delta = \frac{\partial \theta}{\partial t} + \frac{2\pi(\alpha_{22} - \alpha_{11})}{\hbar c} I + \delta\omega \qquad (4)$$

where V_j are the electric field amplitudes of the Raman orders, α_{ij} are the transition moments and $\delta\omega$ is the detuning between the peak frequency separation of the two pump beams and the Raman transition. The term I, is related to the intensity of each MRG order and leads to a Stark shift of each molecular level through a single-color, two-photon interaction. The small difference between the transition moments α_{22} and α_{11}, leads to a change in the Raman transition frequency. It is this Stark shifting that leads the Raman transition to being slightly red-shifted when pumped at high intensity and explained the Sokolov observation [10] of more Raman orders generated with red-detuned pump beams. This Stark shift term needs to be included in the detuning term, Δ. The interaction of the Raman transition with adjacent Raman orders leads to the complex two-photon Rabi frequency $\Omega e^{i\theta}$. A time varying θ gives a frequency detuning term $d\theta/dt$. When using linearly chirped pump pulses, the $d\theta/dt$ term is linearly dependent on both the chirp rate and time delay between the pulses. In Zhi's work [9], a single linearly chirped pump pulse was split into two identical pulses that were then combined with a time delay between the two pulses. In this case, the pump pulses have the same peak frequency making the $\delta\omega$ term equal to the Raman transition frequency. The $d\theta/dt$ from the time delay was used to cancel this detuning. In our experiments, we set the two pump frequencies such that $\delta\omega$ equals zero and the time delay between the linearly chirped pulses detunes the frequencies away from the low intensity resonance. We assume the two pump pulses have the same linear chirp.

Using the two-photon optical Bloch equations with multiple frequency inputs has proven successful in determining the number of Raman orders in both the adiabatic regime [10] and in the transient regime [12]. However, this multi-frequency theory does not allow for any extra peaks to be derived because all possible frequency components, ω_j, are input and only their amplitudes, V_j grow with the Raman gain as a function of propagation length. In order to have extra peaks grow in a theoretical model, the input would have to be a continuum spectrum. Theoretically modeling the observed behavior is beyond the scope of this paper. We do consider what would be expected from a two-photon dressed state picture for the Raman transition. To do this, we will first consider single photon Stark shifting and look at the effect of a weak probe beam tuned through the Stark shifted transition [13]. In his book, Boyd has derived the case for a two-level atom, which is driven by a near resonant, single monochromatic beam. The pump frequency is slightly red-detuned from the transition, that is, the laser frequency is slightly less than the transition. For red-detuning, the population would initially be in the lower coherent state. The calculation of the absorption of a probe beam as its frequency is varied across the transition frequency shows gain on the Stokes side of the transition at a frequency detuning given by the one-photon Rabi frequency and loss on the anti-Stokes side also at the frequency shift given by the Rabi-frequency. The gain on the Stokes side is explained by a three-photon resonance between two manifolds in the dressed state picture from the lower state in the lower manifold to the upper state of the next highest manifold. The loss on the anti-Stokes side is the single photon resonant absorption between two adjacent manifolds also from the lower state to the upper state in the upper manifold. The opposite response would be expected for blue detuning of the pump, as the upper state would now have the initial population and so the transitions would be from the upper state to the lower state of the next highest manifold.

Now if we consider the two-photon dressed states, each of the Raman levels would be split into two coherent states with a separation of Ω', the two-photon Rabi frequency because of strong laser pumping at the pump frequency ω_0 and the first Stokes frequency, ω_{-1}. A schematic of two manifolds in a two-photon dressed state picture is depicted in Figure 1.

The manifolds are separated by the Raman frequency. In order to efficiently generate MRG orders, the pump beams need to be red-detuned to be resonant with the single-color Stark shifted transition. This red-detuning leads to the lower state of each manifold to be the initially populated state. From linear Raman scattering, each of the MRG orders, ω_j, could either amplify a Stokes order ω_{j-1} as depicted in Figure 1a or could as shown in Figure 1c be amplified itself by the scattering of the anti-Stokes order ω_{j+1}. Other resonant interactions are also possible. The system could be left in the higher coherent state and generate a photon at the frequency of $\omega_{j-1} - \Omega'$ as shown in Figure 1b. This Rabi-shifted frequency can then experience gain through stimulated Raman scattering. On the other hand, the higher coherent state can also be reached as depicted in Figure 1d by absorption of the Rabi-shifted anti-Stokes frequency $\omega_{j+1} + \Omega'$ with gain on the MRG order ω_j. This would assume that there were frequencies present at $\omega_{j+1} + \Omega'$, which could be there if a background continuum was generated. In our previous work, where we stretched the pulses to just 600 fs, there was a strong continuum under the Raman peaks and we observed red-shifted shoulders and dips in the continuum spectrum on the blue side of the

Raman orders [7]. Linear Raman scattering of the continuum spectrum can cause gain on the red side of the MRG orders and absorption on the blue side. If the two pump beams were blue detuned from resonance, then the upper dressed state should be in the initially populated state and we would expect gain on the blue shifted side of the peaks and absorption on the red side.t

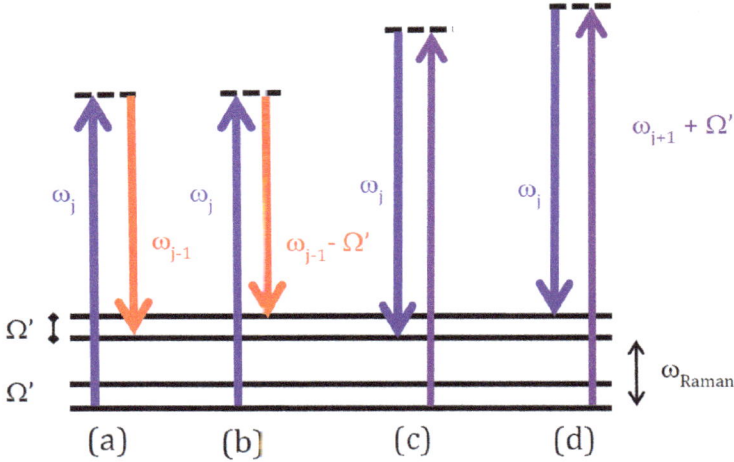

Figure 1. Diagram showing linear Raman scattering from two-photon dressed states; (**a**) scattering of pump beam, ω_j, to Stokes ω_{j-1}; (**b**) scattering of pump beam, ω_j, to Stark shifted Stokes $\omega_{j-1} - \Omega'$; (**c**) gain of pump beam, ω_j by scattering of anti-Stokes beam ω_{j+1}; (**d**) gain of pump beam, ω_j by scattering of Stark shifted anti-Stokes beam $\omega_{j+1} + \Omega'$.

2. Experimental Section

We study MRG in SF_6 because it has a strong Raman vibrational transition and because of its spherical symmetry it does not exhibit rotational lines and so we expect the MRG spectra to be a series of Raman orders separated by a single constant spacing. The Raman frequency in SF_6 is 23.25 Thz, which would give a temporal spacing of ~43 fs in a generated pulse train. The two pump beams are generated in a two-color Ti:sapphire chirped pulse amplification system [14]. The laser is a 10 Hz system that can deliver 400 fs pulses with 1 mJ of energy in each color to the experimental chamber. After the compressor the two colors are collinear. The total energy of the two pumps was varied, by rotating a half-wave plate that was placed before a broadband polarizer. The transmitted p-polarized light was sent to the experimental chamber. The two colors can be individually tuned, the pulse durations are individually compressed and the two pulses are timed by translating the back mirror in one of the compressor lines. To avoid SPM, we leave the pulses with a linear chirp by offsetting the grating separation of the compressor from the optimal position for compression. The pulse duration is typically stretched to ~1 ps with a positive frequency chirp, that is, the red color leads the blue. Because we are using linearly chirped pulses, the instantaneous frequency separation of the two pulses can easily be tuned by varying the time delay between the two linearly chirped pulses.

To determine if the dressed state picture is a reasonable explanation of the red-shifted shoulders, we measured the spectrum as a function of instantaneous frequency separation at three different total pump energies. The experimental MRG system is the same as previously described [8]. We contain the nonlinear medium in a hollow fiber to extend the nonlinear region, guide the beams and provide phase matching. The two-color beam was weakly focused by a 300 mm focal length lens into a 150 μm diameter, 0.5 m long hollow fiber is filled with SF_6, at a pressure 1 atmosphere. We measured the MRG spectra as a function of time delay between the two pump pulses, each having ~4 nm bandwidth and chirped to 1 ps duration, giving a chirp rate of 2 THz/ps. The total maximum pump energy in the two pumps is 2 mJ, with the energy equally split between the two pumps. The two pump frequencies are peaked at frequencies 361 and 384 THz to match the Raman resonance.

3. Results and Discussion

A series of spectra measured at different time delays between the two pump pulses with total energy of 2 mJ, is shown in Figure 2. The spectra were measured with time delay steps of 0.333 ps. The two strong pump beams are not shown on this figure. The lowest frequency line shown is the first anti-Stokes order. The spectral intensity is color coded, such that the lowest intensities are light blue and the most intense is dark red.

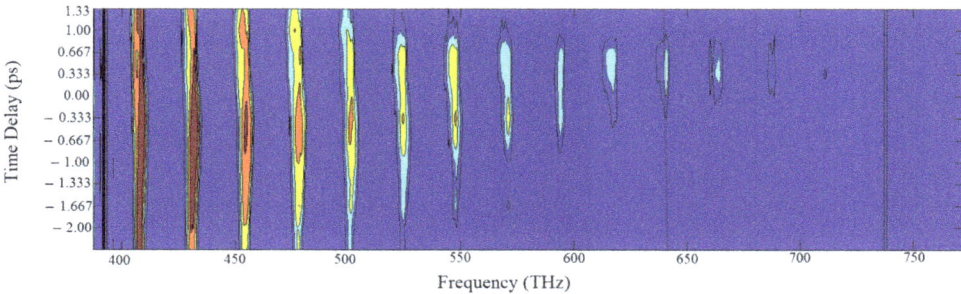

Figure 2. The multi-frequency Raman generation (MRG) spectra are shown for two positively chirped pump pulses with total energy of 2 mJ as a function of time delay between the two pulses.

The spectra shown in Figure 2 display a number of interesting features. First the total number of Raman orders is maximized when a positive time delay of 667 fs is applied. With a delay of two thirds of the pulse duration, the intensity in the pulse overlap region is greatly reduced and yet the nonlinear MRG process has been increased. With a positive delay, the instantaneous frequency separation of the two pumps is less than the Raman frequency. This result of increased orders for a slight red detuning of the pump beams agrees with the results obtained by Sokolov and coworkers, which were taken in the adiabatic regime [10]. It is surprising that pulse durations of just 1 ps would display the same behavior as the adiabatic case. With the time delay increased beyond 1 ps, the number of orders decreases most likely because the intensity of the pulses in the temporal overlap region has further decreased, but it could also indicate that there is an optimum red-detuning. Again as in the adiabatic case, there is no increase in orders for pump frequency separation greater than the Raman transition, which is given in our case for the negative time delays. The decrease in number of orders at increased time delay is more pronounced for negative time delays, indicating that for blue detuning, there is no resonant enhancement to counter the reduced nonlinearity due to the lower overlap intensity.

Secondly, if the red-shifted shoulders were due to non-resonant four-wave mixing, we would expect a plot of MRG spectra as a function of time delay to show the Raman orders as vertical lines and the peaks from non-resonant four-wave mixing would be sloped lines, where the slope would be given by the chirp rate. This is not what we observe. When the time delay is negative, only the vertical lines appear. The spectral orders remain narrow peaks separated by the Raman frequency until zero time delay. At a positive time delay of 667 fs, not only is the number of orders maximized, but also the orders themselves are maximally broadened. The broadening is not linearly dependent on the time delay as would be expected with non-resonant four-wave mixing. The spectral orders do not simply broaden but appear to become double peaked, with a narrow line remaining at the Raman transition and then a broader peak to the red side. The frequency of the red-shifted spectral peak does not vary linearly with pulse delay. We need to look at how a two-photon Rabi frequency shift caused by linearly chirped Gaussian pulses would change with time delay between the two pulses. The Rabi frequency has two terms. The second term is the detuning, Δ, which has three terms: the single color Stark shift which is dependent on the term I, is independent of time delay as is the detuning between the peak frequencies, $\delta\omega$, which in our case is set to zero leaving only the $d\theta/dt$ term which increases linearly in magnitude with time delay. The first term in the Rabi frequency is given by the overlap of the amplitudes $V_j V^*_{j-1}$ which decreases in magnitude with time delay. The decrease is not linear with delay but certainly has the opposite trend to the detuning and so somewhat balances the time dependence of the detuning term. The double peak spectra then agrees with our dressed state picture that allows gain at both the Raman transition and at the two-photon Stark shifted transition, which would have only a small dependence on time delay.

In both the adiabatic and transient experiments enhancement in the number of generated orders only occurs for red-detuning because the single-color Stark shift allows only red-detuned pulses to be resonant with the transition. We have also observed in the transient regime, that there are no blue shifted shoulders with blue detuning indicating that there was no transfer to the coherent Stark shifted states. Pump radiation that is blue-detuned from the low intensity Raman frequency would be further detuned from the Stark shifted transition.

To further confirm that the red-shifted peaks are a result of an intensity dependent Stark shift, we performed the same experiment at lower energies. In Figure 3, we show the time delay spectra for a total energy of 1.5 mJ and in Figure 4, we show the spectra of the 7th to 11th anti-Stokes Raman orders at a time delay of 0.667 ps, for three different pump energies. The total energy of the two pumps was 1.0, 1.5 and 2.0 mJ. The spectra have been normalized so that the pump spectrum would have a peak intensity of 1 on the vertical scale.

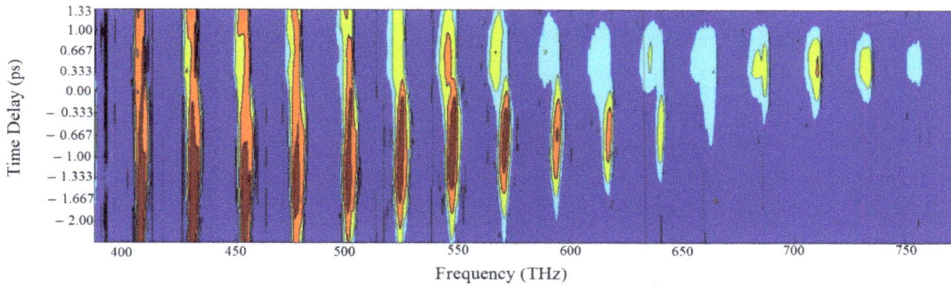

Figure 3. The MRG spectra are shown for two positively chirped pump pulses, with total energy of 1.5 mJ as a function of time delay between the two pulses.

In Figure 3, we can see that the maximum number of orders occurs at the delay of 0.333 ps. This time difference from the 2 mJ result is limited by the resolution of the time step. However it does confirm that at lower intensity, the required red-shift to remain resonant with the single-color Stark shifted transition is reduced.

In Figure 4, the red shifts of the orders are larger with increased pump energy, which is expected if the shifts are due to two-photon Stark shifting. If the extra peaks were due to non-resonant four-wave mixing, the height of the peaks should increase, but not the frequency shift. The peaks that appear at the expected multiples of the Raman transition frequency remain narrow, but the red shifted peaks get broader at the higher orders. At 2 mJ, the average spacing of the red-shifted peaks is 22.7 THz, which lies between the pump frequency separation and the Raman spacing. This could be due to the fact that each order can produce a Stark shifted Stokes frequency, which would result in a small shift from the Raman order. Each of these shifted orders can then undergo Stimulated Raman scattering at either the Raman frequency or the Stark shifted frequency leading to broad peaks after multiple scattering events. This broadening could also be a result of the transient nature of the Stark shifting resulting in different shifts at different intensity points.

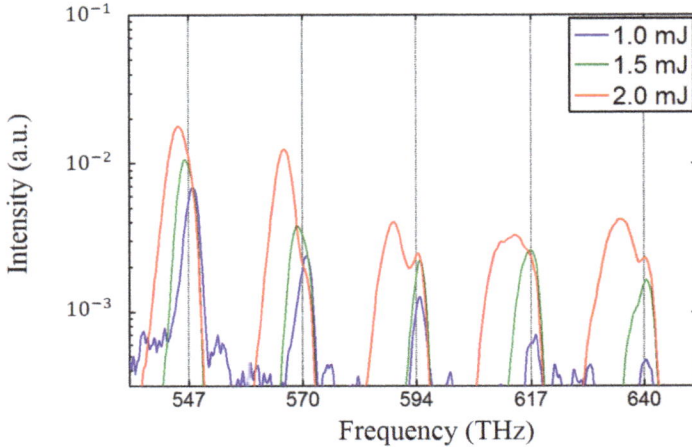

Figure 4. The MRG spectra for the 7th to 11th anti-Stokes orders for three different pump energies. The spectral intensities have arbitrary units, but the plots are normalized to the pump spectrum.

It is surprising that with just 1 ps pulse duration, the results agree with the nanosecond pump duration results, where an adiabatic transition to the coherent state is expected. We measured the MRG spectrum for unchirped pulses that had duration of ~400 fs with zero time delay between the pulses. We tuned the peak frequencies of the laser pulses to vary the frequency separation across the Raman resonance. We show three spectra, each with a total energy of 2 mJ in Figure 5. The top panel shows the spectrum for the two pumps slightly red detuned with a peak separation of 22.5 THz. The middle panel shows the MRG spectrum for on-resonance pumping and the bottom panel shows the spectrum for the blue-detuned pump frequency separation of 24.6 THz. Unlike the linearly chirped pump pulses, the blue shifted pumping displays blue shifted peaks. With on-resonance pumping, the first few orders show red-shifted peaks, but the higher orders appear at the expected Raman frequency. With red-detuned pumps, large red-shifted shoulders appear.

The continuum that appears under the MRG spectra with these short pulses also changes with pump detuning. Although the continuum is dependent on the pump pulse duration, the continuum cannot simply be due to SPM of the pump beams as the largest continuum appears under the higher anti-Stokes Raman orders rather than under the stronger pump frequencies. The continuum appears to be strongest for the red-detuned pumps, indicating that the generation of the continuum is linked with the red-shifted shoulders. The continuum is also centered at higher frequency for the red-detuned pump beams.

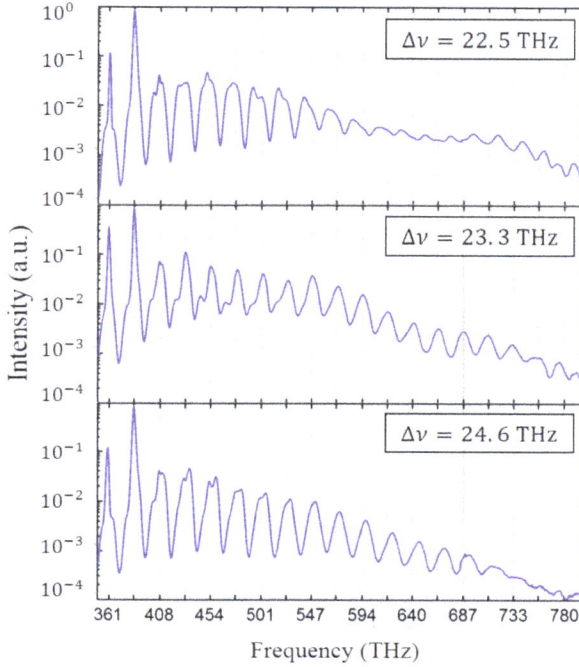

Figure 5. The MRG spectra for 2 mJ, unchirped pump pulses. Pump frequencies are (**top panel**) red detuned by 0.8 THz, (**middle panel**) on-resonance and (**bottom panel**) blue-detuned by 1.3 Thz.

In Figure 6, we have again plotted the 7th through 11th orders for both the red-shifted and blue-shifted short pulse pumping at three different total energies of 1, 1.5 and 2.0 mJ. The short pulse red-detuning of 0.8 THz is less than the red-shift of the chirped pulse at a delay of 0.667 ps, which corresponds to a detuning of 1.3 THz and yet the red-shifting is stronger for each of the pump energies with short pulse pumping compared with the chirped pumping. This is expected from Stark shifting as the pump intensity would be a factor of 2.5 higher for the 400 fs pulses compared to the 1 ps pulses at the same energy. The red-shifted shoulders are also much broader with the short pumps than with the long. This is mostly because the orders extend from the Raman order to the maximum shift, which for 2 mJ, unchirped pulses equals have the transition frequency at the higher orders. The broadening increases with pump energy so that the orders begin to merge and create the continuum that appears under the higher anti-Stckes orders. The fact that the broadening is stronger with the shorter pulses so that the two peaks merge to one broad peak indicates that the broadening is effected by the transient nature of the Stark shifting. This broadening could also come from SPM and cross-phase modulation (XPM) between the orders. One would expect the largest SPM to occur at the pump frequency that has an order of magnitude more intensity and yet the pump frequencies have the narrowest spectra.

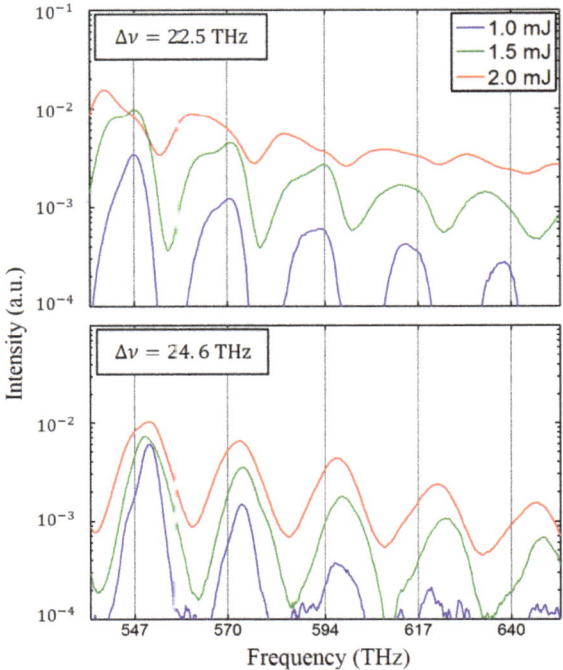

Figure 6. The MRG spectra for the 7th to 11th anti-Stokes orders for three different pump energies. The spectral intensities have arbitrary units, but the plots are normalized to the pump spectrum. The top and bottom plots show the spectra for pump frequency separation of 22.5 and 23.6 THz, respectively. Even so, the red-shift from the Raman orders is larger at each pump energy for the short pulses compared to the chirped pulse results.

From the lower plot in Figure 6, it can be seen that the blue shifted shoulders behave differently from the red-shifting in the upper plot. The position of the blue-shifted peaks is independent of pump energy indicating that the shift is not from the Stark effect. The average spacing of the peaks remains at the low intensity Raman frequency of 23.3 THz. The shifted frequency is stronger than the frequency at the Raman order of the pump frequency. The average shift of the peaks away from the Raman order is 5 THz, which is larger than the 1.3 THz detuning of the pump frequencies. The orders do get broader with intensity leading to a continuum under the orders. This broadening is presumably from SPM and cross-phase modulation between the orders, but still the pump frequencies have the most intensity and narrowest spectra. It is not clear at this point what causes the 5 THz shift or why it is this shifted frequency that then undergoes multiple Raman scattering at the low power Raman transition frequency. The 5 THz shifted frequency could be produced by either higher order non-resonant frequency mixing or by SPM/XPM.

We have not observed blue shifting by time delaying chirped pulses that have peak frequency separation equal to the Raman transition, but we have observed blue shifted shoulders with chirped pulses that had peak frequencies separated by more than the Raman transition [8]. In the chirped case, the shift from the Raman frequency did not increase with order, which is the same as we have now observed with the unchirped pulses. With the chirped pumping the intensity of the blue shifted shoulders remained lower than the orders than appeared at Raman orders of the pump, which is different than what we observe with unchirped pulses. Whereas the red-shifting could occur by either time detuning chirped pulses or by detuning the peak frequencies, the blue shift requires that the peak frequencies be blue detuned, whether the pumps are chirped or not. This dependence on the peak frequencies suggests that the blue shift is due to a non-resonant frequency mixing process. Since the single-color Stark effect causes a red-detuning of the transition, this leaves the blue detuned pulses even further from the Raman resonance and that would allow the non-resonant frequency mixing of the peak frequencies to dominate.

4. Conclusions

We have demonstrated that when using pump pulses in the transient regime, extra peaks occur in the multi-frequency Raman generated spectra. These peaks occur only on the red side of the Raman orders when pumped with linearly chirped pump pulses. The frequency shift of the red-detuned peaks does not follow the instantaneous frequency separation of the pump pulses indicating that non-resonant four-wave mixing is not responsible for the red-shifted peaks. On the other hand, the amount of shift is intensity dependent, which implies that the extra peaks are a result of Stark shifting of the Raman levels. We noted that pumping with chirped pulse duration of just 1 ps in the transient regime gives very similar results to the adiabatic regime where the molecular states are expected to evolve into the coherent states. In particular, more Raman orders are generated when the instantaneous frequency separation of the pump beams are slightly red-shifted from the Raman transition, but blue shifting does not change the spectrum from the on-resonant pumping. Linear Raman scattering from a two-photon Rabi frequency shifted coherent states can explain the generation of the red-shifted shoulders that appear in the MRG spectra. In comparison with chirped laser pumping, unchirped pulse pumping with detuned peak frequencies displays very different behavior. In particular, not only are the orders red-shifted for red-detuned pumps, but the orders are blue shifted when the pumps are blue shifted, indicating that non-resonant frequency mixing becomes a dominant process when the pulses are unchirped. Although a two-photon Stark shift can explain the red-shifted shoulders and also why blue shifting of the pumps allows a non-resonant process to dominate, it remains to be determined why a constant blue shift for all orders is observed with blue detuned peak pump frequencies.

Acknowledgments: The authors gratefully acknowledge funding support from the Natural Sciences and Engineering Research Council.

Conflicts of Interest: The authors declare no conflict of interest.

References

1. Imasaka, T.; Kawasaki, S.; Ishibashi, N. Generation of more than 40 laser emission lines from the ultraviolet to the visible regions by two-color stimulated raman effect. *Appl. Phys. B* **1989**, *49*, 389–392.
2. Shverdin, M.Y.; Walker, D.R.; Yavuz, D.D.; Yin, G.Y.; Harris, S.E. Generation of a single-cycle optical pulse. *Phys. Rev. Lett.* **2005**, *94*, 033904.
3. Yoshii, K.; Anthony, J.K.; Katsuragawa, M. The simplest route to generating a train of attosecond pulses. *Light: Sci. Appl.* **2013**, *2*, e58.
4. Losev, L.L.; Lutsenko, A.P. Ultrabroadband parametric stimulated Raman scattering in a highly transient regime. *Opt. Commun.* **1996**, *132*, 489–493.
5. Sali, E.; Mendham, K.J.; Tisch, J.W.G.; Halfmann, T.; Marangos, J.P. High-order stimulated Raman scattering in a highly transient regime driven by a pair of ultrashort pulses. *Opt. Lett.* **2004**, *29*, 495–497.
6. Turner, F.C.; Trottier, A.; Losev, L.L.; Strickland, D. Transient multi-frequency Raman generation in SF$_6$. *Opt. Commun.* **2007**, *270*, 419–423.
7. Turner, F.C.; Strickland, D. Anti-stokes enhancement of multi-frequency Raman generation in a hollow fibre. *Opt. Lett.* **2008**, *33*, 405–407.
8. Cui, Z.; Chaturvedi, M.; Tian, B.; Ackert, J.; Turner, F.C.; Strickland, D. Spectral red-shifting of multi-frequency Raman orders. *Opt. Commun.* **2013**, *288*, 118–121.
9. Zhi, M.; Sokolov, A.V. Broadband coherent generation in a Raman crystal driven by a pair of time-delayed linearly chirped pulses. *New J. Phys.* **2008**, *10*, 025032.
10. Sokolov, A.V.; Walker, D.R.; Yavuz, D.D.; Yin, G.Y.; Harris, S.E. Raman generation by phased and antiphased molecular states. *Phys. Rev. Lett.* **2000**, *85*, 562–565.
11. Hickman, A.P.; Paisner, J.A.; Bischel, W.K. Theory of multiwave propagation and frequency conversion in a Raman medium. *Phys. Rev. A* **1986**, *33*, 1788–1797.

12. Sali, E.; Kinsler, P.; New, G.H.C.; Mendham, K.J.; Halfmann, T.; Tisch, J.W.G.; Marangos, J.P. Behavior of high-order stimulated Raman scattering in a highly transient regime. *Phys. Rev. A* **2005**, *72*, 013813.

13. Boyd, R.W. Nonlinear optics in the two-level approximation. In *Nonlinear Optics*, 3rd ed.; Elsevier: Oxford, UK, 2008; pp. 277–328.

14. Xia, J.F.; Song, J.; Strickland, D. Development of a dual-wavelength Ti:sapphire multi-pass amplifier and its application to intense mid-infrared generation. *Opt. Commun.* **2002**, *206*, 149–157.

Continuous-Wave Molecular Modulation Using a High-Finesse Cavity

David C. Gold, Joshua J. Weber and Deniz D. Yavuz

Abstract: We demonstrate an optical modulator at a frequency of 90 THz that has the capability to modulate any laser beam in the optical region of the spectrum. The modulator is constructed by placing deuterium molecules inside a high-finesse cavity and driving a vibrational transition with two continuous-wave laser beams. The two beams, the pump and the Stokes, are resonant with the cavity. The high intra-cavity intensities that build up drive the molecules to a coherent state. This molecular coherence can then be used to modulate an independent laser beam, to produce frequency up-shifted and down-shifted sidebands. The beam to be modulated is not resonant with the cavity and thus the sidebands are produced in a single pass.

Reprinted from *Appl. Sci.* Cite as: Gold, D.C.; Weber, J.J.; Yavuz, D.D. Continuous-Wave Molecular Modulation Using a High-Finesse Cavity. *Appl. Sci.* **2014**, *4*, 498–514.

1. Introduction

Over the last two decades, we have witnessed an explosion of interest and progress in ultrafast science [1]. Femtosecond time scale molecular dynamics (vibrational and rotational) are now routinely probed using commercially available laser systems. The advances in femtosecond lasers have also allowed the generation of soft X-ray subfemtosecond pulses using the technique of high harmonic generation (HHG) [2,3]. The frontiers of ultrafast science have now moved comfortably into the subfemtosecond domain and electronic processes on these time scales are currently being studied. Despite this great progress, many researchers (including us) feel that key ingredients of ultrafast physics remain missing. We cannot yet produce optical waveforms with the same precision and flexibility with which we can produce electronic waveforms. We also have not yet been able to produce a coherent optical spectrum with a large number of precise laser beams that simultaneously cover the infrared, visible, and ultraviolet regions of the electromagnetic spectrum. Constructing such a device, typically referred to as an arbitrary optical waveform generator [4], has been one of the biggest unmet challenges in the field since the invention of the laser in 1960. This device would ideally produce fully coherent radiation ranging from 200 nanometers all the way up to 10 microns in wavelength (a "white-light" laser).

We believe that, if constructed, such a device would have important applications in a broad range of research areas, such as the following examples. (i) By phase locking the spectral components, one can synthesize subfemtosecond pulses in the optical region of the spectrum [5–8]. Although, as mentioned above, subfemtosecond pulses are now routinely produced in the soft X-ray region of the spectrum using HHG, synthesizing such pulses in the more accessible optical region will increase their utility and impact. (ii) By appropriately adjusting the phases and the amplitudes of the Fourier components, one can synthesize arbitrary optical waveforms, such as a square waveform or a sawtooth waveform, on subfemtosecond time scales. Such sub-cycle pulse shaping will likely extend the horizons of coherent and quantum control to a completely new regime [9–11]. (iii) A continuous-wave (CW) white-light laser with narrow linewidth spectral components would allow precision spectroscopy of a large number of molecular and atomic species [12]. Furthermore, the whole spectrum could be locked to a reference so that the absolute frequency of each component is known to a high precision. Using this approach, it may be possible to construct optical clocks in different regions of the spectrum or to generate a broad absolute frequency reference with components covering the full optical region.

One of the most promising techniques for the construction of an arbitrary optical waveform generator is the technique of *molecular modulation* [13–17]. This technique utilizes coherent vibrations and rotations in molecules to produce a broad Raman spectrum covering many octaves of bandwidth. A subset of such a spectrum has recently been used to synthesize optical waveforms with sub-cycle resolution [18–20]. In this technique, the molecules are driven to a highly coherent state using two intense laser beams, the pump and the Stokes. Because of the high intensities required for efficient modulation, experiments have traditionally been performed using Q-switched pulsed lasers. The extension of the molecular modulation technique to the CW domain would allow for each spectral component to have a very narrow frequency linewidth. Such precision is required in spectroscopy and many other applications. Over the last five years, we have been working towards accomplishing this goal. To achieve the intensities required by the molecular modulation technique with CW lasers, our approach has been to place the molecules inside a cavity with a high finesse at the pump and at the Stokes wavelengths [21–27]. Our prior experiments have demonstrated CW stimulated Raman scattering in a previously inaccessible regime: at very low gas pressures with an established CW molecular coherence more than two orders of magnitude larger than previously achieved [28]. Of particular importance, we have demonstrated for the first time that this approach can produce a broad spectrum in wavelength regions where the cavity mirrors are not reflective [29–31].

In this paper, building on our earlier work, we report two experimental results. (i) We have constructed an optical modulator that can modulate any laser in the optical region of the spectrum with a modulation frequency of 90 THz. The modulator is prepared by driving a vibrational transition in molecular D_2 using the resonant pump and the Stokes beams in the cavity. With the molecules coherently prepared, a separate laser beam passes through the molecular gas cell and is modulated in a single pass. (ii) We have achieved locking of two independent laser beams to the cavity (the pump and the Stokes) and observed Raman generation as we varied the two-photon detuning. This is the first experiment where a CW molecular modulator is prepared using two independent laser beams. These results show considerable promise for fully extending the molecular modulation technique to the CW domain. Specifically, they open up the prospect of using a molecular modulator to broaden the spectrum of a broadband laser (such as a Ti:sapphire femtosecond oscillator), which we will discuss below.

2. Molecular Modulation

The technique of molecular modulation differs from traditional stimulated Raman scattering [32–37] in that it relies on Raman generation in the regime of near maximal coherence. Figure 1a shows a simplified energy level diagram for a typical molecular system (for example, D_2). Here state $|a\rangle$ is the ground rotational-vibrational state of the molecule and state $|b\rangle$ is a particular excited rotational-vibrational state. The two excitation lasers (the pump and the Stokes) drive the coherence (off-diagonal density matrix element ρ_{ab}) of the Raman transition. When the driving lasers are sufficiently intense, it becomes possible to coherently transfer almost half of the population to the excited state, thus approaching the state of maximum coherence, $|\rho_{ab}| \approx 1/2$. The concept of maximum coherence has gained considerable attention over the last two decades. It is now understood that for nonlinear mixing processes, the generation efficiency is maximized for a maximally coherent state [38–41]. Furthermore, because the generation process is almost complete in a single coherence length, phase-matching is unimportant and new beams are generated collinearly.

When the molecules are driven to a highly coherent state, they act as an efficient frequency mixer with the modulation frequency determined by the oscillation frequency of the molecules, $\nu_M = \nu_{pump} - \nu_{Stokes}$. For linearly polarized driving laser beams, the driving lasers themselves can be modulated to produce higher-order Stokes and anti-Stokes sidebands through four-wave mixing. In addition to the driving lasers, the system can also be used to modulate any other laser beam. As shown in Figure 1, a separate laser (called the mixing beam) with frequency ν_0 can be modulated to produce frequency upshifted (anti-Stokes) and downshifted (Stokes) sidebands at frequencies $\nu_0 + \nu_M$ and $\nu_0 - \nu_M$, respectively. If the process is highly efficient or a multi-pass configuration is used, a very broad Raman spectrum can be produced with frequencies of the form $\nu_0 + q\nu_M$ (q is an integer). Using the molecular modulation technique with pulsed lasers, several groups have demonstrated the generation of a wide spectrum in molecular H_2 and D_2 [14–17]

104

and have synthesized the first optical pulses to break the single-cycle barrier [18–20]. These experiments currently hold the world record for the shortest pulses ever synthesized in the optical region of the spectrum. When the driving laser beams are opposite circularly polarized, as is the case in Figure 1a, the generation of sidebands from the driving lasers themselves is suppressed due to angular momentum conservation rules [42]. In this regime, a frequency upshifted or downshifted sideband from an appropriately-polarized mixing beam will be produced, and the molecular medium can be used as an efficient frequency converter. Further details regarding the molecular modulation technique can be found in a number of review articles [43–45].

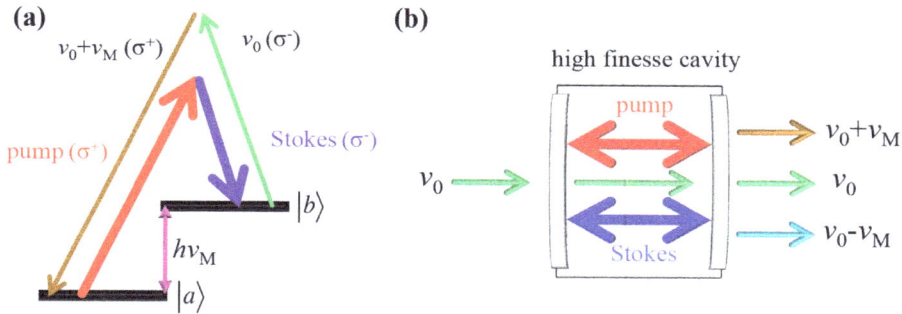

Figure 1. (a) Simplified energy level diagram. Two intense lasers, the pump and the Stokes, drive the Raman transition and establish the molecular coherence, ρ_{ab}. The molecules can then be used to modulate any other mixing laser with frequency ν_0 to produce frequency upshifted or downshifted sidebands at frequencies $\nu_0 + \nu_M$ and $\nu_0 - \nu_M$, respectively. With the polarization configuration as shown, only a frequency upshifted sideband is produced. (b) Molecular modulation with CW laser beams. The molecules are placed inside a cavity with a high finesse at the pump and at the Stokes wavelengths. For a sufficiently large molecular coherence, any mixing laser incident on the cavity will be modulated. The mixing beam does not need to be resonant with the cavity as the modulation is produced in a single pass through the system.

2.1. Molecular Modulation with CW Lasers

Despite these exciting experiments, molecular modulation technique as demonstrated in these experiments has significant limitations and has not yet made a big impact on other research areas. The required laser intensities for preparing the molecules into a maximally coherent state is quite high, exceeding $100 \, \text{MW/cm}^2$. As a result, all of those molecular modulation experiments described in the previous paragraph have been performed using Q-switched pulsed lasers, which have important limitations. The duty cycle of these lasers is low, less than

1 part in 10^7, which severely restricts data rates. Furthermore, each sideband has a minimum linewidth that is determined by the duration of the Q-switched pulse, $1/(10 \text{ ns}) \approx 100$ MHz. This is an important limitation for precision spectroscopy experiments where the transitions studied may have linewidths on the Hz level. To overcome these limitations, it is important to extend the molecular modulation technique to the CW domain.

For extending the technique to the CW domain, two approaches are promising. One approach is to utilize the tight confinement of hollow-core photonic crystal fibers, an approach that has been studied by Benabid and colleagues [46]. In contrast, our approach is to place the molecules inside a cavity with high finesse at the pump and at the Stokes wavelengths, as shown in Figure 1b. When the cavity resonance condition is simultaneously satisfied for both laser beams, high intensities build up inside the cavity and the molecules are driven to a highly coherent state. If the pump and the Stokes beams are linearly polarized, higher-order Stokes and anti-Stokes beams from the driving lasers will be produced. A separate weaker mixing laser can also be modulated in a single pass through the system. Our scheme was motivated by the recent pioneering experiments of Carlsten and colleagues that have demonstrated CW Raman lasing of the Stokes beam inside a high-finesse cavity [21–26,35]. We have extended these experiments to the regime of sufficiently high coherence such that significant modulation can be produced in a single pass through the system. As a result, the mixing beam does not need to be resonant with the cavity, and any optical wavelength can be modulated. This broadband modulation capability is critical, as it may allow for the broadening of the already broad spectrum of a Ti:sapphire oscillator.

3. Experiment: 90 THz CW Modulator

In this section, we discuss our optical modulator with a modulation frequency of 90 THz that is capable of modulating any beam in the optical region of the spectrum. Optical modulators typically utilize nonlinear optical processes in crystals, such as electro-optic and acousto-optic effects. Such modulators have recently achieved modulation rates approaching 100 GHz; yet, due to intrinsic limitations of electronic components, crystal-based approaches may remain unable to achieve much higher modulation rates. In contrast, we have recently demonstrated a modulator at a frequency of 17.6 THz using a rotational transition in molecular H_2 [30]. The experiment that we describe in this section extends our earlier result to a higher modulation rate of 90 THz by using a vibrational transition in molecular D_2.

A schematic of our experiment is shown in Figure 2. We prepare the molecules to a highly coherent state inside a high-finesse cavity with two intense laser fields, the pump and the Stokes, at wavelengths of 1.064 μm and 1.555 μm, respectively. The experiment starts with locking the pump laser beam to one of the cavity modes. For sufficiently high intra-cavity pump laser intensity, the Stokes laser beam is produced through Raman lasing into a cavity mode. We utilize the $|v = 0, J = 0\rangle \rightarrow |v = 1, J = 0\rangle$ vibrational Raman transition in D_2, which has a transition frequency of 90 THz. With the molecules prepared in a coherent state, a third, weaker laser beam at a wavelength of 785 nm passes through the system. This mixing beam is modulated to produce frequency up-shifted and down-shifted sidebands at wavelengths of 636 nm and 1026 nm, respectively. The mirrors of the cavity do not have a high reflectivity at 785 nm, thus the modulation is produced in a single pass through the system.

To produce the desired high power pump laser beam, we start with a custom-built external cavity diode laser (ECDL) with an optical power of 20 mW and a free-running linewidth of about 0.5 MHz. We amplify the ECDL output with an ytterbium fiber amplifier centered at 1.064 µm. The amplifier produces a linearly-polarized, single-spatial and single frequency mode output with a maximum power of 20 W. The amplified beam is then coupled to the TEM_{00} mode of the high-finesse cavity using a mode matching lens (MML). The mirrors of the high-finesse cavity have high damage threshold coatings with high reflectivity near wavelengths of 1.06 µm and 1.55 µm. The transmittance of the mirrors at the two wavelengths are about 50 parts per million (ppm) and the total scattering and absorption losses are at the level of 100 ppm. One of the cavity mirrors is mounted on a piezoelectric transducer to allow for slight adjustments of the cavity length. We use the Pound–Drever–Hall (PDH) technique to lock the amplifier output to the cavity [47]. For this purpose, the ECDL passes through an electro-optic modulator (EOM) and the cavity reflected signal is picked off with a glass slide (GS) to produce electronic feedback to the ECDL frequency and the cavity length. The mirrors are housed in a vacuum chamber, which is machined from stainless steel with an inner tube diameter of 5 cm. The central 50 cm-long region of the chamber is surrounded by a liquid N_2 reservoir. Cooling the molecules reduces Doppler broadening and greatly increases the population of the ground rotational level [14]. The mixing beam is produced by a separate 785 nm ECDL whose output is amplified by a semiconductor tapered amplifier. The optical power in the mixing beam before the cavity is about 100 mW. The motivation for amplifying the mixing beam is to increase the optical power of the generated sidebands to more easily detectable levels.

Figure 3 shows the power conversion efficiency from the mixing beam to the frequency up-shifted sideband at a wavelength of 636 nm, as the incident pump power is varied. For this measurement, the D_2 pressure is held constant at 0.3 atm. The highest conversion efficiency that we achieve is 0.5×10^{-6}. Although the conversion efficiency is low, this is the first time CW modulation of an independent laser at such a high rate has been demonstrated. In addition to the measurement of Figure 3, as we vary the incident pump power, we also measure the transmitted pump and Stokes powers through the cavity. Similar to our previous experiments, the transmitted pump power stays constant at about 1 mW, whereas the transmitted Stokes power increases linearly to a maximum of about 7 mW as we increase the input pump power. From the transmitted power measurements, we can calculate the intra-cavity circulating intensities for the pump and the Stokes beams using the mirror transmittance and the known cavity mode size. The maximum calculated intra-cavity intensities for the pump and Stokes lasers are 10 kW/cm^2 and 62 kW/cm^2, respectively. The solid line in Figure 3 shows the calculated conversion efficiency using the estimated intra-cavity pump and Stokes intensities. There are no adjustable parameters in this calculation; *i.e.*, all parameters that are used are experimentally measured. We observe reasonable agreement between the experimental data points and the calculations. The disagreement is likely due to the imperfect spatial overlap of the 785-nm beam with the TEM$_{00}$ mode of the cavity. This spatial overlap is difficult to measure precisely in our experiment and the calculations do not take this effect into account. The predicted largest value for the molecular coherence that we achieve in this experiment is $|\rho_{ab}| = 2.5 \times 10^{-5}$.

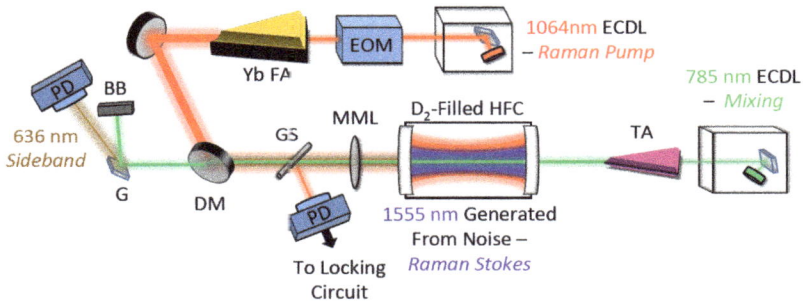

Figure 2. The simplified schematic of our experiment. We start the experiment by locking a high-power pump laser beam to one of the axial modes of the cavity. With the pump laser locked, the Stokes beam is produced through Raman lasing. The pump and the Stokes beams drive the molecular coherence, which can then be used to modulate a separate 785-nm laser beam in a single pass. ECDL: external cavity diode laser, EOM: electro-optic modulator, Yb FA: ytterbium fiber amplifier, PBS: polarizing beam splitter, GS: glass slide, PD: photo-diode, MML: mode-matching lens, HFC: high-finesse cavity, TA: tapered amplifier, DM: dichroic mirror, G: grating, BB: beam block.

Currently, the conversion efficiency in this experiment is limited by the intra-cavity pump and Stokes laser intensities. One drawback of the current experiment is that we rely on Raman lasing to generate the Stokes beam. As a result, the pump and the Stokes intra-cavity intensities are limited by the dynamics of Raman lasing. This drawback can be overcome by locking two independent laser beams to the cavity, rather than relying on Raman lasing, an approach we discuss in the next section.

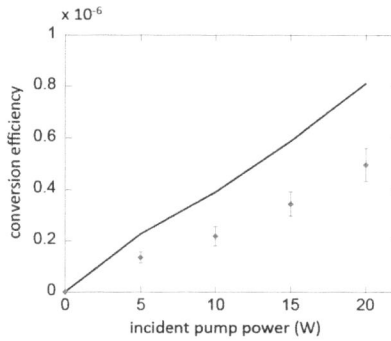

Figure 3. The conversion efficiency from the mixing beam to the 636 nm anti-Stokes sideband as the incident pump power is varied at a constant D_2 pressure of 0.3 atm. The solid line is a theoretical calculation without any adjustable parameters, based on the measured transmitted powers of the pump and the Stokes beams through the cavity.

4. Experiment: Preparing Molecules with Two Independent CW Laser Beams

In this section, we discuss our experiment where we have prepared the molecular coherence by using two independent laser beams. Figure 4 shows the simplified schematic of this experiment. Similar to the pump laser, the Stokes laser system starts with an ECDL, but now near a wavelength of 1.55 μm. We amplify the Stokes ECDL output to 20 W with an erbium-doped fiber amplifier. A PDH setup similar to that of the pump laser keeps the Stokes beam resonant with the cavity. When we lock both lasers to the cavity and adjust the frequency difference of the two lasers to within about a GHz of the Raman transition resonance, $\nu_{pump} - \nu_{Stokes} \approx \nu_{ab}$, the two beams efficiently drive the molecular coherence. The established molecular coherence then mixes with the driving lasers to produce the anti-Stokes and second Stokes sidebands at wavelengths of 807 nm and 2.94 μm, respectively. These sidebands are only observed when the lasers are tuned to within about 1 GHz of the Raman transition. For these experiments, we reduce the gas pressure to a sufficiently low value such that Raman lasing on the vibrational Stokes beam is negligible; *i.e.*, if we lock only the pump laser to the cavity there is no substantial Raman generation. The Raman lasing needs to be suppressed because the Stokes beam generated through Raman lasing interferes with the cavity lock of the 1.55 μm beam. Figure 5 shows the power of the generated 807 nm anti-Stokes beam as a function of the two-photon detuning at a molecular gas pressure of 0.01 atm. This pressure value is about three orders of magnitude lower than those traditionally used in stimulated Raman scattering experiments. It is important to note that, just as in the 90 THz mixing experiment described above, the cavity mirrors do not have a high reflectivity at the anti-Stokes wavelength, so the 807 nm beam is generated in a single pass through the system. We generate a maximum CW anti-Stokes power exceeding 1 mW, which is an order of magnitude larger than our previous experiments that relied on Raman lasing to generate the Stokes beam [30]. The predicted established molecular coherence in this experiment is more than an order of magnitude higher than what was achieved in the mixing experiment, $|\rho_{ab}| \approx 1 \times 10^{-3}$. Although the established molecular coherence in this experiment is much higher, we have not yet been able to translate this increase into an increase in the mixing efficiency experiment of the previous section. When we increase the molecular gas pressure, the dynamics of Raman lasing and instabilities in the locking performance substantially reduce the intra-cavity intensities for the pump and the Stokes beams.

One of our immediate goals is to increase the established molecular coherence and operate at higher pressures in the two-beam setup and thereby increase the mixing efficiency. We believe that the intra-cavity intensities in our experiment are currently limited by the large free-running linewidths of the ECDLs (about 0.5 MHz) and by imperfect locking electronics. Due to these limitations, we are able to couple at best 10% of the incident power in each laser beam to the cavity. To overcome these limitations, we plan to pursue the following technical improvements in our setup in the near future: (i) constructing better ECDLs with free-running linewidths of about 50 kHz [48]; (ii) pre-stabilizing the ECDLs with separate low-finesse cavities so that their free-running linewidths are considerably reduced [49]; and (iii) increasing the bandwidth of the locking electronics and better understanding the limitations of the feedback circuit [50,51]. With these improvements, our goal will be to achieve intra-cavity intensities approaching the damage threshold of the mirror coatings (about 100 MW/cm^2). CW optical intensities exceeding 100 MW/cm^2 inside a high-finesse cavity have recently been experimentally demonstrated for high-quality dielectric coatings [52]. Our experience with these coatings has been consistent with this recent experiment since we have been able to routinely obtain intensities exceeding 10 MW/cm^2 in an empty chamber (without any molecular gas) without any apparent degradation in mirror performance. With both the pump and the Stokes beams locked to the cavity with circulating intensities of about 100 MW/cm^2, the established molecular coherence will be $|\rho_{ab}| \approx 0.01$, more than an order of magnitude higher than what we have achieved to date. We can then use the system to modulate a separate weak mixing laser to produce frequency upshifted and downshifted sidebands. We would then have a 90 THz modulator with a single-pass conversion efficiency approaching 10% over much of the optical region of the spectrum. As described in the next section, such an efficient broadband modulator has exciting potential applications.

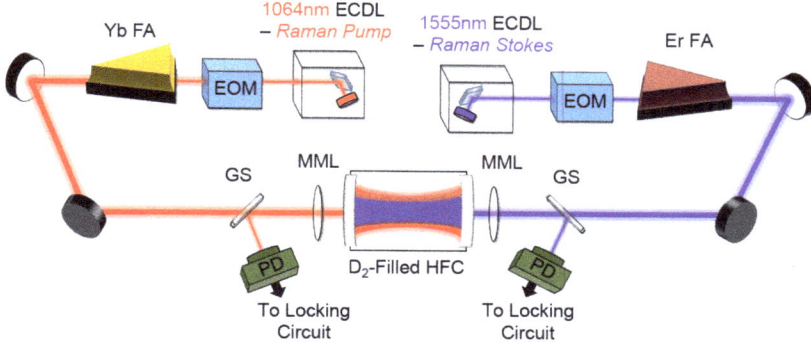

Figure 4. Simplified schematic for the two-beam experiment. We start with two external cavity diode lasers at the pump and at the Stokes wavelengths. Each beam is amplified to an output power of 20 W using a fiber amplifier. The amplified beams are then coupled to the high-finesse cavity using mode matching lenses. We use the Pound–Drever–Hall technique to lock the amplifier outputs to the cavity. ECDL: external cavity diode laser, EOM: electro-optic modulator, Yb FA: ytterbium-doped fiber amplifier, Er FA: erbium-doped fiber amplifier, GS: glass slide, MML: mode matching lens, PD: photo-diode, HFC: high-finesse cavity.

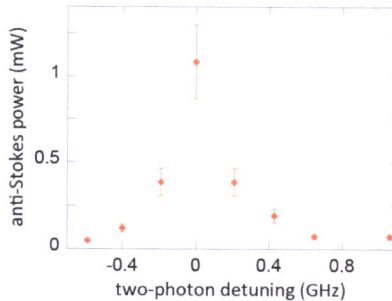

Figure 5. The power of the 807 nm anti-Stokes beam as a function of two-photon Raman detuning, $(\nu_{pump} - \nu_{Stokes}) - \nu_{ab}$, at a D_2 gas pressure of 0.01 atm. At these low pressures, Doppler broadening is the dominant contribution to the linewidth. The two-photon detuning is varied by locking either the pump or the Stokes beam to a different cavity mode. As a result, the data points are spaced in frequency by the free spectral range of the cavity, which is 200 MHz.

5. Modulation of a Femtosecond Ti:sapphire Oscillator

One of the main motivations for extending molecular modulation to the CW domain is that it would allow combining this technique with femtosecond Ti:sapphire laser technology. The idea is that the modulator can be used to efficiently generate sidebands on a femtosecond oscillator to produce a broad spectrum with a large number of spectral components. This scheme was first suggested by Kien *et al.* [53], and was later experimentally demonstrated with Q-switched pulsed lasers by Marangos and colleagues [54–56]. Consider a broadband laser with power spectral density $F(\nu)$. If this laser goes through a molecular medium that is prepared in a highly coherent state, modulation produces sideband "replicas" of the spectrum spaced from the original by the modulation frequency. The first Stokes and anti-Stokes sidebands will produce replicas $F(\nu - \nu_M)$ and $F(\nu + \nu_M)$, the second orders will produce $F(\nu - 2\nu_M)$ and $F(\nu + 2\nu_M)$, and this process could continue until the spectrum fills the full optical range. Figure 6 shows how the scheme works for a typical broadband Ti:sapphire laser oscillator with a spectrum that spans from about 700 nm to 900 nm in wavelength. In this qualitative calculation, we use the vibrational transition in D_2 ($\nu_M = 90$ THz) to produce sidebands and broaden the Ti:sapphire spectrum. To show the effect more clearly, we assume 60% conversion efficiency to each sideband, although as mentioned above, in the near future, we will aim for about 10% conversion efficiency with $|\rho_{ab}| \approx 0.01$. Intensity spectra (in linear scale) for the incident Ti:sapphire beam and for first, second, and third order modulation are shown.

A key advantage of this approach, and the feature that sets it apart from similar previous work, is that since the molecules are prepared with two CW lasers, the linewidth of the molecular oscillations can be made very narrow. When the two driving lasers are locked to the high-finesse cavity, their linewidths can be narrowed to values much smaller than 1 kHz. This also sets the linewidth of the molecular oscillations (*i.e.*, the precision in the value of ν_M). As a result, the mixing process will not significantly affect the spectral characteristics of the individual comb lines of the Ti:sapphire oscillator. Many of the ideas and experiments that use stable Ti:sapphire comb lines may be performed with the much broader spectrum of Figure 6. We also note that the broad output spectrum of Figure 6 can form a phase-stabilized frequency comb. This can be accomplished by adjusting the modulation frequency ν_M to be an exact integer multiple of the comb-spacing of the Ti:sapphire laser. The modulation frequency is precisely set by the frequencies of the pump and the Stokes laser beams, $\nu_M = \nu_{pump} - \nu_{Stokes}$. The pump and Stokes laser beams can, for example, be locked to two teeth of a separate phase-stabilized frequency-comb oscillator.

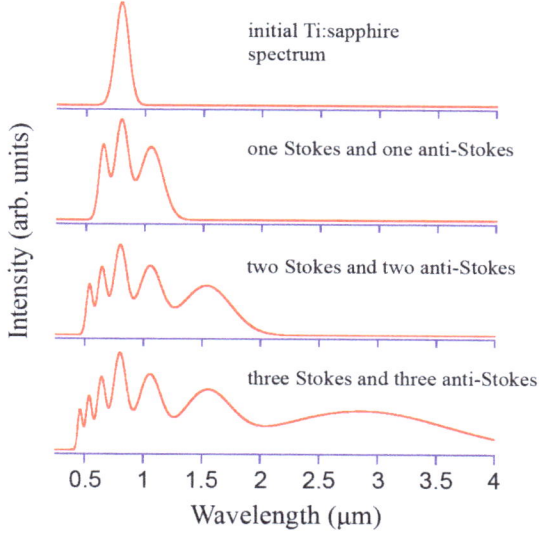

Figure 6. Broadening the spectrum of a Ti:sapphire laser using the vibrational transition in D_2 with $\nu_M = 90$ THz. Starting with a typical broadband laser with power spectral density, $F(\nu)$, replicas of the spectrum spaced by the Raman transition frequency are produced. Here we follow the process up to three Stokes and anti-Stokes sidebands, *i.e.*, the final spectrum includes contributions from $F(\nu - 3\nu_M)$, $F(\nu - 2\nu_M)$, ..., $F(\nu + 3\nu_M)$. In this calculation, we assume 60% conversion efficiency to each sideband.

We note that throughout this section, we have assumed the Ti:sapphire laser to be sufficiently weak such that it does not interfere with the molecular state preparation. This assumption is satisfied for a typical femtosecond oscillator (1 nanojoule energy per pulse, 20 femtosecond pulse width, 100 MHz repetition rate). For many applications such as quantum control, nanojoule per pulse level energies will not be sufficient. For such applications, an amplified Ti:sapphire system (larger than 1 microjoule energy per pulse, about 10 kHz repetition rate) can be used as a mixing laser. The peak power of such lasers is quite high, and these pulses would interfere with the molecular state preparation by the pump and the Stokes laser beams. To overcome this problem, one idea would be to chirp the pulse so that its peak power is greatly reduced. In the spirit of the technique of *chirped pulse amplification*, peak power can be dramatically reduced by stretching the pulses to picosecond times scales. Also, for modulating amplified Ti:sapphire systems, pulsed molecular modulators such as those utilizing photonic crystal fibers may be used [57].

We conclude this section by noting that there are other schemes that can be used to broaden the spectrum of a Ti:sapphire laser. Early experiments used nonlinear self-phase modulation inside a photonic crystal fiber to broaden the Ti:sapphire spectrum [58,59]. Recently, multi-octave broadband spectra and single-cycle pulse synthesis have been demonstrated using a number of approaches including coherently combining the supercontinuum output from multiple highly nonlinear fibers and using multiple optical parametric amplifiers seeded by a common laser source [5–8]. These are exciting developments that may eventually allow for the synthesis of arbitrary optical waveforms. However, we believe that the simplicity and the coherent nature of the molecular modulation process make it the most promising route for constructing an arbitrary optical waveform generator. As mentioned before, the molecular modulation technique was the first to synthesize single-cycle pulses and this technique still holds the world record for the shortest pulses that have ever been synthesized in the optical region of the spectrum [18–20].

6. Conclusions and Future Directions

In conclusion, we have discussed constructing an optical modulator at a frequency of 90 THz that has the capability of modulating any laser in the optical region of the spectrum. We also have discussed the initial results of our efforts to prepare the molecular modulator using two independent lasers locked to the modes of a high-finesse cavity. These results are significant steps towards extending the molecular modulation technique to the CW domain. The long-term goal of our project is to use the molecules to frequency broaden the spectrum of a Ti:sapphire oscillator as shown in Figure 6. If successful, this approach may produce a broad CW spectrum covering the full optical region with millions of spectral components. We next discuss some of the exciting potential applications of such a source.

6.1. Temporal Waveform Synthesis

Once a very broad spectrum with millions of spectral components is produced, the next challenge will be to synthesize temporal waveforms using pulse-shaping techniques [60,61]. The full optical spectrum (from 200 nm to 10 microns) is much broader than the typical spectra used in pulse-shaping experiments. Because of the large bandwidth of the spectrum, arbitrary temporal waveform synthesis will likely be a challenging task, and there are a number of open questions. To adjust the phases and the amplitudes of components over such a broad spectrum, one approach would be to use piezo-driven moveable metal mirrors [62–64] and to work with a reflective geometry. These mirrors have a high reflection coefficient over the full optical region, and recent advances in fabrication allow for a large number of individually adjustable pixels. For the infrared region of the spectrum where the wavelength is large, the limited range of piezo-motion might be an issue. To overcome this limitation, an acousto-optic pulse shaper may be used for the infrared wavelengths [65]. A key component of pulse-shaping techniques is a

114

nonlinear detector that allows measurement of the synthesized waveform through correlation studies. This is traditionally a thin nonlinear crystal where the second harmonic signal as a function of time delay reveals the temporal structure. However, for a very broad spectrum, crystals cannot be used; a nonlinear detector with a wide bandwidth is required. The nonlinear processes in noble gases have a sufficiently large bandwidth for this purpose. For example, the experiments of the Harris group used multi-photon ionization and sum frequency generation in Xe atoms to characterize ultrafast pulses [18].

6.2. Frequency Reference and Precision Time-Keeping

There is a growing demand for constructing clocks with improved precision. Recent advances in optical clocks have yielded frequency stabilities exceeding 1 part in 10^{15}, surpassing the precision of Cs atomic clocks [66–68]. Optical clocks utilize ultra-stable lasers that are locked to very narrow atomic transition lines. With a CW molecular modulator, it may be possible to extend the frequency stability of an optical clock to the full optical spectrum. As previously mentioned, a key advantage is that since the molecules are prepared with two CW lasers, the linewidth of the molecular oscillations can be made very narrow. The frequencies of the two driving laser beams, ν_{pump} and ν_{Stokes}, can in principle be locked to optical reference lines (such as to two teeth of a stabilized, frequency-comb Ti:sapphire laser). As a result, the absolute frequency of the molecular oscillations, $\nu_M = \nu_{pump} - \nu_{Stokes}$, would be known to a very high precision (potentially approaching a precision at the level of 1 part in 10^{15}). The idea then would be to use a frequency stabilized mixing beam so that all mixing orders with frequencies $\nu_0 + q\nu_M$ have an absolute frequency stability on the 1 part in 10^{15} level. As discussed above, we could also use a Ti:sapphire laser as a mixing beam. For this case, if the Ti:sapphire laser is absolute frequency stabilized (a frequency comb), the generated spectrum of Figure 6 will be an absolute frequency reference covering the full optical region. We believe this would constitute a light source with many exciting applications in a number of research areas. For example, one can perform precision spectroscopy of a large number of atomic and molecular species from a single source. One would also be able to construct optical clocks in regions of the spectrum that are currently inaccessible.

6.3. Coherent Control

Over the last decade, quantum and coherent control has emerged as an exciting sub-field of atomic, molecular, and optical physics [69–71]. It is now understood that efficient control of a quantum system requires laser light with a broad spectrum with frequencies that match as many transitions in the system as possible [72]. Using arbitrary optical waveforms with a spectrum ranging from 200 nm to 10 μm in wavelength, one can simultaneously control the electronic, vibrational, and rotational coordinates of molecules. This will facilitate the understanding of various couplings between different coordinates of the system. For example, one possibility is to study the ionization of a molecule as a function of the vibrational coordinate. This cannot be done with only the soft X-ray pulses provided by HHG since their spectrum does not have the infrared wavelengths that are needed to efficiently control the slow vibrational coordinate.

The ability to shape the pulse beyond the slowly-varying envelope approximation will likely extend the ideas of coherent control to a completely new regime. Currently, because of the limited bandwidth of laser sources, coherent control experiments are done in the quasi-steady-state regime, in which the electric field of the laser light cycles back and forth many times before its envelope changes. Sub-cycle pulse shaping allows true manipulation of the electric field itself instead of just the envelope, providing a completely new tool. Up until now, sub-cycle pulses have only been available in the THz region of the spectrum. These pulses have been used for many pioneering experiments in Rydberg atoms in which the Rydberg wave-packets are precisely controlled [73,74]. Extending sub-cycle pulses to the optical region of the spectrum will extend these experiments to electronic wave-packets with much shorter characteristic time scales. As demonstrated by the recent exciting experiments from the HHG community [75], there remains much to be understood about the ultrafast dynamics of electronic motion and its coupling to other coordinates.

Acknowledgments: We thank Tyler Green for his contributions to the early stages of the project. We also thank Nick Proite, Dan Sikes, Zach Simmons, Jared Miles, and Nick Brewer for many helpful discussions. Since 2009, this work has been supported by the National Science Foundation (NSF). We also acknowledge support from the Wisconsin Alumni Research Foundation (WARF).

Author Contributions: D. D. Yavuz supervised the research, designed the experiments, and wrote the text. D. C. Gold and J. J. Weber performed the experiments, took the data, and edited the text. All authors contributed to the interpretation of the experimental results.

Conflicts of Interest: The authors declare no conflict of interest.

Bibliography

1. Brabec, T.; Krausz, F. Intense few-cycle laser fields: frontiers of nonlinear optics. *Rev. Mod. Phys.* **2000**, *72*, 545–591.
2. Bucksbaum, P.H. The future of attosecond spectroscopy. *Science* **2007**, *317*, 766–769.
3. Kapteyn, H.; Cohen, O.; Christov, I.; Murnane, M. Harnessing attosecond science in the quest for coherent X-rays. *Science* **2007**, *317*, 775–778.
4. Yavuz, D.D. Toward synthesis of arbitrary optical waveforms. *Science* **2011**, *331*, 1142–1143.

5. Krauss, G.; Lohss, S.; Hanke, T.; Sell, A.; Eggert, S.; Huber, R.; Leitenstorfer, A. Synthesis of a single cycle of light with compact erbium-doped fibre technology. *Nat. Photonics* **2010**, *4*, 33–36.

6. Ycas, G.; Osterman, S.; Diddams, S.A. Generation of a 660–2100 nm laser frequency comb based on an erbium fiber laser. *Opt. Lett.* **2012**, *37*, 2199–2201.

7. Wirth, A.; Hassan, M.T.; Grguras, I.; Gagnon, J.; Moulet, A.; Luu, T.T.; Pabst, S.; Santra, R.; Alahmed, Z.A.; Azzeer, A.M.; *et al.* Synthesized light transients. *Science* **2011**, *334*, 195–200.

8. Huang, S.W.; Cirmi, G.; Moses, J.; Hong, K.H.; Bhardwaj, S.; Birge, J.R.; Chen, L.J.; Li, E.; Eggleton, B.J.; Cerullo, G.; *et al.* High-energy pulse synthesis with sub-cycle waveform control for strong-field physics. *Nat. Photonics* **2011**, *5*, 475–479.

9. Judson, R.S.; Rabitz, H. Teaching lasers to control molecules. *Phys. Rev. Lett.* **1992**, *68*, 1500–1503.

10. Baumert, T.; Brixner, T.; Seyfried, V.; Strehle, M.; Gerber, G. Femtosecond pulse shaping by an evolutionary algorithm with feedback. *Appl. Phys. B* **1997**, *65*, 779–782.

11. Zare, R.N. Laser control of chemical reactions. *Science* **1998**, *279*, 1875–1879.

12. Hansch, T.W.; Walther, H. Laser spectroscopy and quantum optics. *Rev. Mod. Phys.* **1999**, *71*, S242–S252.

13. Harris, S.E.; Sokolov, A.V. Broadband spectral generation with refractive index control. *Phys. Rev. A* **1997**, *55*, R4019–R4022.

14. Sokolov, A.V.; Walker, D.R.; Yavuz, D.D.; Yin, G.Y.; Harris, S.E. Raman generation by phased and anti-phased molecular states. *Phys. Rev. Lett.* **2000**, *85*, 562–565.

15. Lian, J.Q.; Katsuragawa, M.; Kien, F.L.; Hakuta, K. Sideband generation using strongly driven Raman coherence in solid hydrogen. *Phys. Rev. Lett.* **2000**, *85*, 2474–2477.

16. Katsuragawa, M.; Liang, J.Q.; Le Kien, F.; Hakuta, K. Efficient frequency conversion of incoherent fluorescent light. *Phys. Rev. A* **2002**, *65*, 025801, doi:10.1103/PhysRevA.65.025801.

17. Yavuz, D.D.; Walker, D.R.; Shverdin, M.Y.; Yin, G.Y.; Harris, S.E. Quasi-periodic Raman technique for ultrashort pulse generation. *Phys. Rev. Lett.* **2003**, *91*, 233602, doi:10.1103/PhysRevLett.91.233602.

18. Shverdin, M.Y.; Walker, D.R.; Yavuz, D.D.; Yin, G.Y.; Harris, S.E. Generation of a single-cycle optical pulse. *Phys. Rev. Lett.* **2005**, *94*, 033904, doi:10.1103/PhysRevLett.94.033904.

19. Chen, W.; Hsieh, Z.; Huang, S.; Su, H.; Tang, T.; Lin, C.; Lee, C.; Pan, R.; Pan, C.; Kung, A.H. Sub-single-cycle optical pulse train with constant carrier envelope phase. *Phys. Rev. Lett.* **2008**, *100*, 163906, doi:10.1103/PhysRevLett.100.163906.

20. Chan, H.S.; Hsieh, Z.M.; Liang, W.H.; Kung, A.H.; Lee, C.K.; Lai, C.J.; Pan, R.P.; Peng, L.H. Synthesis and measurement of ultrafast waveforms from five discrete optical harmonics. *Science* **2011**, *331*, 1165–1168.

21. Brasseur, J.K.; Repasky, K.S.; Carlsten, J.L. Continuous-wave Raman laser in H_2. *Opt. Lett.* **1998**, *23*, 367–369.

22. Brasseur, J.K.; Roos, P.A.; Repasky, K.S.; Carlsten, J.L. Characterization of a continuous-wave Raman laser in H_2. *J. Opt. Soc. Am. B* **1999**, *16*, 1305–1312.

23. Roos, P.A.; Meng, L.S.; Carlsten, J.L. Using an injection-locked diode laser to pump a CW Raman paser. *IEEE J. Quant. Electron.* **2000**, *36*, 1280–1283.

24. Shinzen, K.; Hirakawa, Y.; Imasaka, T. Generation of highly repetitive optical pulses based on intracavity four-wave Raman mixing. *Phys. Rev. Lett.* **2001**, *87*, 223901, doi:10.1103/PhysRevLett.87.223901.

25. Ihara, K.; Eshima, C.; Zaitsu, S.; Kamitomo, S.; Shinzen, K.; Hirakawa, Y.; Imasaka, T. Molecular-optic modulator. *Appl. Phys. Lett.* **2006**, *88*, 074101, doi:10.1063/1.2174091.

26. Zaitsu, S.; Eshima, C.; Ihara, K.; Imasaka, T. Generation of a continuous-wave pulse train at a repetition rate of 17.6 THz. *J. Opt. Soc. Am. B* **2007**, *24*, 1037–1041.

27. Zaitsu, S.; Imasaka, T. Phase-matched generation of high-order continuous-wave coherent Raman sidebands. *Opt. Commun.* **2012**, *285*, 347–351.

28. Green, J.T.; Sikes, D.E.; Yavuz, D.D. Continuous-wave, high-power rotational Raman generation in molecular deuterium. *Opt. Lett.* **2009**, *34*, 2563–2565.

29. Green, J.T.; Weber, J.J.; Yavuz, D.D. Continuous-wave light modulation at molecular frequencies. *Phys. Rev. A* **2010**, *82*, 011805(R), doi:10.1103/PhysRevA.82.011805.

30. Weber, J.J.; Green, J.T.; Yavuz, D.D. 17 THz continuous-wave optical modulator. *Phys. Rev. A* **2012**, *85*, 013805, doi:10.1103/PhysRevA.85.013805.

31. Weber, J.J.; Yavuz, D.D. Broadband spectrum generation using continuous-wave Raman scattering. *Opt. Lett.* **2013**, *38*, 2449–2451.

32. Yoshikawa, S.; Imasaka, T. A new approach for the generation of ultrashort optical pulses. *Opt. Commun.* **1993**, *96*, 94–98.

33. Kaplan, A.E. Subfemtosecond pulses in mode-locked 2π solitons of the cascade stimulated Raman scattering. *Phys. Rev. Lett.* **1994**, *73*, 1243–1246.

34. McDonald, G.S.; New, G.H.C.; Losev, L.L.; Lutsenko, A.P.; Shaw, M. Ultrabroad-bandwidth multifrequency Raman generation. *Opt. Lett.* **1994**, *19*, 1400–1402.

35. Kawano, H.; Hirakawa, Y.; Imasaka, T. Generation of more than 40 rotational Raman lines by picosecond and femtosecond Ti:sapphire laser for Fourier synthesis. *Appl. Phys. B* **1996**, *65*, 1–4.

36. Kawano, H.; Hirakawa, Y.; Imasaka, T. Generation of high-order rotational lines in hydrogen by four-wave Raman mixing in the femtosecond regime. *IEEE J. Quantum Electron.* **1998**, *34*, 260–268.

37. Nazarkin, A.; Korn, G.; Wittmann, M.; Elsaesser, T. Generation of multiple phase-locked Stokes and anti-Stokes components in an impulsively excited Raman medium. *Phys. Rev. Lett.* **1999**, *83*, 2560–2563.

38. Scully, M.O.; Zubairy, M.S. *Quantum Optics*; Cambridge University Press: Cambridge, UK, 1997.

39. Harris, S.E. Electromagnetically Induced Transparency. *Phys. Today* **1997**, *50*, 36–42.

40. Jain, M.; Xia, H.; Yin, G.Y.; Merriam, A.J.; Harris, S.E. Efficient Nonlinear frequency conversion with maximal atomic coherence. *Phys. Rev. Lett.* **1996**, *77*, 4326–4329.

41. Merriam, A.J.; Sharpe, S.J.; Xia, H.; Manuszak, D.; Yin, G.Y.; Harris, S.E. Efficient gas-phase generation of coherent vacuum ultraviolet radiation. *Opt. Lett.* **1999**, *24*, 625–627.

42. Sokolov, A.V.; Sharpe, S.; Shverdin, M.Y.; Walker, D.R.; Yavuz, D.D.; Yin, G.Y.; Harris, S.E. Optical frequency conversion by a rotating molecular waveplate. *Opt. Lett.* **2001**, *26*, 728–730.

43. Walker, D.R.; Yavuz, D.D.; Shverdin, M.Y.; Yin, G.Y.; Harris, S.E. A Quasi-periodic approach for femtosecond pulse generation. *Opt. Photonics News* **2003**, *47*, 46–51.

44. Sokolov, A.V.; Shverdin, M.Y.; Walker, D.R.; Yavuz, D.D.; Burzo, A.M.; Yin, G.Y.; Harris, S.E. Generation and control of femtosecond pulses by molecular modulation. *J. Mod. Opt.* **2005**, *52*, 285–304.

45. Shverdin, M.Y.; Walker, D.R.; Yavuz, D.D.; Goda, S.; Yin, G.Y.; Harris, S.E. Breaking the single-cycle barrier. *Photonics Spectra* **2005**, *39*, 92–105.

46. Couny, F.; Benabid, F.; Light, P.S. Subwatt threshold CW Raman fiber-gas laser based on H_2-filled hollow-core photonic crystal fiber. *Phys. Rev. Lett.* **2007**, *99*, 143903, doi:10.1103/PhysRevLett.99.143903.

47. Drever, R.W.P.; Hall, J.L.; Kowalski, F.V.; Hough, J.; Ford, G.M.; Munley, A.J.; Ward, H. Laser phase and frequency stabilization using an optical resonator. *Appl. Phys. B* **1983**, *31*, 97–105.

48. Baillard, X.; Gauguet, A.; Bize, S.; Lemonde, P.; Laurent, Ph.; Clairon, A.; Rosenbusch, P. Interference-filter-stabilized external-cavity diode lasers. *Opt. Commun.* **2006**, *266*, 609–613.

49. Ludlow, A.D.; Huang, X.; Notcutt, M.; Zanon-Willette, T.; Foreman, S.M.; Boyd, M.M.; Blatt, S.; Ye, J. Compact, thermal-noise-limited optical cavity for diode laser stabilization at 1×10^{-15}. *Opt. Lett.* **2007**, *32*, 641–643.

50. Roos, P.A.; Brasseur, J.K.; Carlsten, J.L. Intensity-dependent refractive index in a nonresonant CW Raman laser that is due to thermal heating of the Raman active gas. *J. Opt. Soc. Am. B* **2000**, *17*, 758–763.

51. Biefang, C.J.; Rudolph, W.; Roos, P.A.; Meng, L.S.; Carlsten, J.L. Steady-State Thermo-optic model of a continuous-wave Raman laser. *J. Opt. Soc. Am. B* **2002**, *19*, 1318–1325.

52. Meng, L.S.; Brasseur, J.K.; Neumann, D.K. Damage threshold and surface distortion measurement for high-reflectance, low-loss mirrors to 100+ MW/cm^2 CW laser intensity. *Opt. Express* **2005**, *13*, 10085–10091.

53. Kien, F.L.; Hong, S.N.; Hakuta, K. Generation of subfemtosecond pulses by beating a femtosecond pulse with a Raman coherence adiabatically prepared in solid hydrogen. *Phys. Rev. A* **2001**, *64*, 051803, doi:10.1103/PhysRevA.64.051803.

54. Gundry, S.; Anscombe, M.P.; Abdulla, A.M.; Sali, E.; Tisch, J.W.G.; Kinsler, P.; New, G.H.C.; Marangos, J.P. Ultrashort-pulse modulation in adiabatically prepared Raman media. *Opt. Lett.* **2005**, *30*, 180–182.

55. Gundry, S.; Anscombe, M.P.; Abdulla, A.M.; Hogan, S.D.; Sali, E.; Tisch, J.W.G.; Marangos, J.P. Off-resonant preparation of a vibrational coherence for enhanced stimulated Raman scattering. *Phys. Rev. A* **2005**, *72*, 033824, doi:10.1103/PhysRevA.72.033824.

56. Bustard, P.J.; Sussman, B.J.; Walmsley, I.A. Amplification of impulsively excited molecular rotational coherence. *Phys. Rev. Lett.* **2010**, *104*, 193902, doi:10.1103/PhysRevLett.104.193902.

57. Abdolvand, A.; Walser, A.M.; Ziemienczuk, M.; Nguyen, T.; Russel, P.S.J. Generation of a phase locked Raman frequency comb in gas-filled hollow-core photonic crystal fiber. *Opt. Lett.* **2012**, *37*, 4362–4364.

58. Ranka, R.W.J.; Stentz, A. Visible continuum generation in air-silica microstructure optical fibers with anolamous dispersion. *Opt. Lett.* **2000**, *25*, 25–27.

59. Diddams, S.A.; Jones, D.J.; Ma, L.S.; Cundiff, C.T.; Hall, J.L. Optical frequency measurement across a 104-THz gap with a femtosecond laser frequency comb. *Opt. Lett.* **2000**, *25*, 186–188.

60. Hillegas, C.W.; Tull, J.X.; Goswami, D.; Strickland, D.; Warren, W.S. Femtosecond laser pulse shaping by use of microsecond radio-frequency pulses. *Opt. Lett.* **1994**, *19*, 737–739.

61. Weiner, A.M. Femtosecond optical pulse shaping and processing. *Prog. Quantum Electron.* **1995**, *19*, 161–237.

62. Zeek, E.; Maginnis, K.; Backus, S.; Russek, U.; Murnane, M.; Mourou, G.; Kapteyn, H.; Vdovin, G. Pulse compression by use of deformable mirrors. *Opt. Lett.* **1999**, *24*, 493–495.

63. Hacker, M.; Stobrawa, G.; Sauerbrey, R.; Buckup, T.; Motzkus, M.; Wildenhain, M.; Gehner, A. Micromirror SLM for femtosecond pulse shaping in the ultraviolet. *Appl. Phys. B* **2003**, *B76*, 711–714.

64. Radzewicz, C.; Wasylczyk, P.; Wasilewski, W.; Krasinski, J.S. Piezo-driven deformable mirror for femtosecond pulse shaping. *Opt. Lett.* **2004**, *29*, 177–179.

65. Strasfeld, D.B.; Shim, S.; Zanni, M.T. Controlling vibrational excitation with shaped mid-IR pulses. *Phys. Rev. Lett.* **2007**, *99*, 038102, doi:10.1103/PhysRevLett.99.038102.

66. Diddams, S.A.; Berquist, J.C.; Jefferts, S.R.; Oates, C.W. Standards of time and frequency at the outset of the 21st century. *Science* **2004**, *306*, 1318–1324.

67. Ludlow, A.D.; Boyd, M.M.; Zelevinsky, T.; Foreman, S.M.; Blatt, S.; Notcutt, M.; Ye, J. Systematic study of the [87]Sr clock transition in an optical lattice. *Phys. Rev. Lett.* **2006**, *96*, 033003, doi:10.1103/PhysRevLett.96.033003.

68. Takamoto, M.; Hong, F.; Higashi, R.; Katori, H. An optical lattice clock. *Nature* **2005**, *435*, 321–324.

69. Warren, W.S.; Rabitz, H.; Dahleh, M. Coherent control of quantum dynamics: The dream is alive. *Science* **1993**, *259*, 1581–1589.

70. Brixner, T.; Krampert, G.; Pfeifer, T.; Selle, R.; Gerber, G.; Wollenhaupt, M.; Graefe, O.; Horn, C.; Liese, D.; Baumer, T. Quantum control by ultrafast polarization shaping. *Phys. Rev. Lett.* **2004**, *92*, 208301, doi:10.1103/PhysRevLett.92.208301.

71. Bartels, R.A.; Weinacht, T.C.; Leone, S.R.; Kapteyn, H.C.; Murnane, M.M. Nonresonant control of multimode molecular wave-packets at room temperature. *Phys. Rev. Lett.* **2002**, *88*, 033001, doi:10.1103/PhysRevLett.88.033001.

72. Shapiro, M.; Brumer, P. Coherent control of molecular dynamics. *Rep. Prog. Phys.* **2003**, *66*, 859, doi:10.1088/0034-4885/66/6/201.

73. Ahn, J.; Hutchinson, D.N.; Rangan, C.; Bucksbaum, P.H. Quantum phase retrieval of a Rydberg wave packet using a half-cycle pulse. *Phys. Rev. Lett.* **2001**, *86*, 1179–1182.

74. Murray, J.M.; Pisharody, S.N.; Wen, H.; Rangan, C.; Bucksbaum, P.H. Information hiding and retrieval in Rydberg wave packets using half-cycle pulses. *Phys. Rev. A* **2006**, *74*, 043402, doi:10.1103/PhysRevA.74.043402.

75. Baltuska, A.; Udem, T.; Uiberacker, M.; Hentschel, M.; Goulielmakis, E.; Gohle, Ch.; Holzwarth, R.; Yakovlev, V.S.; Scrinzi, A.; Hansch, T.W.; *et al.* Attosecond control of electronic processes by intense light fields. *Nature* **2003**, *421*, 611–614.

Broadband Continuous-Wave Multi-Harmonic Optical Comb Based on a Frequency Division-by-Three Optical Parametric Oscillator

Yen-Yin Lin, Po-Shu Wu, Hsiu-Ru Yang, Jow-Tsong Shy and A. H. Kung

Abstract: We report a multi-watt broadband continuous-wave multi-harmonic optical comb based on a frequency division-by-three singly-resonant optical parametric oscillator. This cw optical comb is frequency-stabilized with the help of a beat signal derived from the signal and frequency-doubled idler waves. The measured frequency fluctuation in one standard deviation is ~437 kHz. This is comparable to the linewidth of the pump laser which is a master-oscillator seeded Yb:doped fiber amplifier at ~1064 nm. The measured powers of the fundamental wave and the harmonic waves up to the 6th harmonic wave are 1.64 W, 0.77 W, 3.9 W, 0.78 W, 0.17 W, and 0.11 W, respectively. The total spectral width covered by this multi-harmonic comb is ~470 THz. When properly phased, this multi-harmonic optical comb can be expected to produce by Fourier synthesis a light source consisting of periodic optical field waveforms that have an envelope full-width at half-maximum of 1.59 fs in each period.

Reprinted from *Appl. Sci.* Cite as: Lin, Y.-Y.; Wu, P.-S.; Yang, H.-R.; Shy, J.-T.; Kung, A.H. Broadband Continuous-Wave Multi-Harmonic Optical Comb Based on a Frequency Division-by-Three Optical Parametric Oscillator. *Appl. Sci.* **2014**, *4*, 515–524.

1. Introduction

The development of ultrafast physics has made tremendous progress in the past decade. This includes the development of broadband coherent optical sources which are desirable in ultrafast source development as well as in a wide range of research areas including laser spectroscopy and quantum optics [1]. While it is possible to produce ultrashort pulses via Fourier waveform synthesizer at optical frequencies that leads to a periodic train of single-cycle attosecond pulses [2–4], more attention has been given to isolated pulse generations in the femtosecond and attosecond time domain. This is because an isolated pulse can be employed readily for experiments in probing the electronic dynamic in atoms and molecules [5]. The common approaches that have been utilized for generating broadband multi-octave coherent optical sources include self-phase-modulation/cross phase-modulation of femtosecond laser pulses in gases and solids [6,7], driving Raman resonances to produce sidebands by four-wave mixing with nanosecond lasers or femtosecond lasers [8–10], synchronously pumped femtosecond optical parametric oscillators [11] or supercontinuum seeded optical parametric amplifiers [12]. A common feature of these approaches is that they use a high power pulsed laser as the pump source. Although there are impressive demonstrations of using these sources as femtosecond waveform synthesizers [13,14], stringent requirement on phase stability dictates that the synthesized waveform easily varies with time unless elaborate feedback control is installed in the system [13].

The opposite extreme to isolated pulses is a continuous train of sub-cycle optical pulses. Since cw lasers are intrinsicaly more stable and are better managed than pulsed lasers, a train of sub-cycle pulses at ~100 THz can offer unprecedented precision and accuracy to as much as one part in 10^{14} in ultrafast event timing, ranging, quantum control, and for metrologic applications. A Fourier waveform synthesizer using a phase-locked cw laser has a main attribute that pulsed lasers do not have. It can maintain highly precise phase and amplitude stability for each comb component to yield waveforms that are nearly free from waveform variation [3]. A cw periodic waveform can therefore be used to produce ultra-stable novel shaped field potentials for trapping neutrals and charged particles and study their short-range dynamical behavior.

There are several new developments toward a broadband coherent optical source for the purpose of making a cw Fourier waveform synthesizer. A cavity-enhanced molecular modulation scheme has been demonstrated [15]. The spectral components of the molecular modulator were extended to span from 0.8 μm to 3.2 μm via four-wave mixing in a gas medium. A 4-THz cw frequency comb at 1.56 μm based on cascading quadratic nonlinearities has also been realized [16]. Yet the power of the frequency bands produced in these developments are in the mW level so that their potential applications are limited. Here we describe an approach that utilizes reliable high power cw fiber lasers. The approach employs a master-oscillator fiber-laser power amplifier pumped frequency division-by-three parametric oscillator [17,18] to provide the first three comb components and then adopt quasi-phase-matched (QPM) nonlinear mixing to generate the next three higher harmonics to give a highly-stable commensurate six component cw comb at the watt level. The approach we describe here encompasses advantages of QPM based parametric oscillators and frequency conversions which are compact and have inherently high conversion efficiency.

2. Experimental Section

The schematic of our multi-harmonic optical comb is shown in Figure 1. The watt-level frequency division-by-three singly resonant optical parametric oscillator (SRO) cavity is a bow-tie ring cavity consisting of two curved mirrors (M1 and M2) with a 100 mm radius of curvature, a flat mirror (M3), and an output coupler (M4).

The OPO crystal is a 5 mol % MgO-doped PPLN crystal (from HC Photonics) with a length of 40 mm. The total cavity length is 500 mm. The pump, signal, and idler waves are designed to correspond to the 3rd harmonic wave, the 2nd harmonic wave, and the fundamental wave of a multi-harmonic optical comb. M1 and M2 have reflectance of <2%, >99.9%, and <5% at the pump, signal and idler wavelengths, respectively. M3 has reflectance of >99.9% at the signal wavelength. Output coupler mirror M4 has a 0.6% output coupling at the signal wavelength.

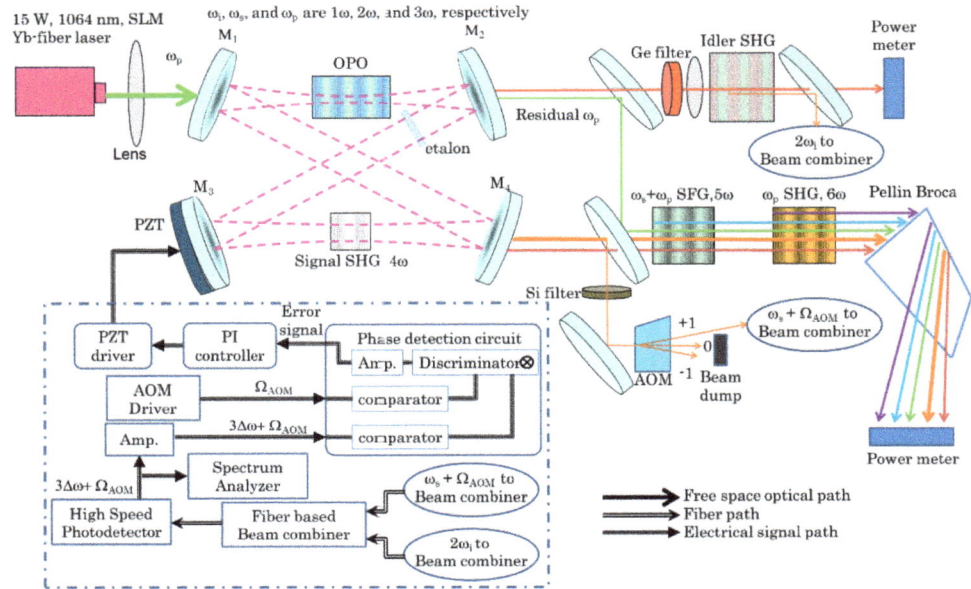

Figure 1. Multi-harmonic optical comb based on a frequency division-by-three singly resonant optical parametric oscillator. The optical parametric oscillator generates the fundamental wave (idler), the 2nd harmonic wave (signal), and the 3rd harmonic wave (pump) of the multi-harmonic optical comb. The 4th harmonic to 6th harmonic are generated by intracavity second harmonic generation (SHG), external cavity single-pass sum frequency generation (SFG), and external cavity single-pass SHG, respectively. The idler (the fundamental wave) SHG and a small portion of the 2nd harmonic wave are combined to produce an error signal for frequency control by tweaking the cavity length that is based on a home-made phase detection circuit, a commercial high-speed proportional integrating (PI) controller (New Focus LB1005) and a high voltage PZT driver (Physik Instruments E-501.621) shown inside the dashed box

The two end faces of the MgO:PPLN crystal are optically polished and antireflection-coated with <1%, <0.2%, and <5% reflectance at pump, signal, and idler wavelengths. The OPO crystal has a period of 30.8 μm and generates radiation at 1596-nm (signal) and at 3192-nm (idler) at 92.8 °C when pumped at 1064 nm. The pump laser is a cw linearly polarized, single-frequency Yb-doped fiber laser amplifier which is seeded by a single-mode mW-level DFB diode laser that has <0.1 MHz linewidth at 1064 nm from IPG Photonics. The pump laser is capable of producing ~15-W power with TEM$_{00}$ mode and a M^2 < 1.05. The cavity is designed to resonate at the signal wave that has a focusing parameter ξ = L/b of 1 (L is the crystal length and b is the confocal parameter). The pump beam is focused to a beam waist radius of 60 μm to mode-match to the cavity mode of the signal beam. The pump and idler exit the cavity from M2 and the signal output is at M4. The pump and idler are separated by a dichroic mirror for their power measurements. The idler wave is immediately focused into a 25 mm long MgO:PPLN crystal of period 34.25 μm for SHG of the idler to produce a beat signal with the signal wave. The purpose of the beat signal is to lock the length of the SRO cavity. This beat signal which is detected by a fast photodiode (EOT ET-3010) is sent to a cavity stabilization feedback circuit shown enclosed in the dash box in Figure 1. Locking the cavity length in turn fixes the wavelength of the signal to one of the cavity modes. A 250-μm thick intracavity fused silica etalon is inserted between M2 and M3 to maintain single-longitudinal-mode operation at high pump power.

The 4th harmonic to 6th harmonic waves can be generated by either intracavity or external cavity configurations. Intracavity frequency conversion could provide higher conversion efficiency, but the dynamic of the SRO cavity becomes more complicated and the cavity mirrors' coatings are more susceptible to damage. So only the 4th harmonic comb component is generated by intracavity SHG from the 2nd harmonic wave (SRO signal wave). The nonlinear crystal for 4th harmonic wave generation is a 20.58 μm period MgO:PPLN crystal with a 10-mm length. The 5th and 6th harmonic waves are generated by single pass sum-frequency generation (SFG) of the 2nd harmonic wave and the 3rd harmonic wave (SRO signal and pump), and SHG of the 3rd harmonic wave (SRO pump) respectively. The specifications of these two nonlinear crystals are a 12.05 μm period MgO:PPLN crystal with 25 mm in length and a 7.97 μm period MgO:PPSLT crystal with 30 mm in length, respectively.

3. Results and Discussion

3.1. Frequency Control of Division-by-Three OPO

When the optical parametric oscillator is operating, the frequencies of signal and idler are $2\omega + \Delta\omega$ and $\omega - \Delta\omega$ where $\Delta\omega$ is the deviation in frequency from the fundamental (idler) of the exact division-by-three frequency of the SRO where the pump is at 3ω. Ideally $\Delta\omega$ is to be equal to zero. For the SRO, zeroing $\Delta\omega$ is done by adjusting the temperature of the gain crystal, the intracavity etalon, and the length of the cavity. The primary source of non-zero $\Delta\omega$ is thermal and mechanical fluctuation of the SRO cavity length. Therefore it is necessary to track and stabilize the cavity length. Here a cavity stabilization feedback circuit is introduced to control the free spectral range of the SRO cavity to eliminate or minimize the frequency deviation. In our experiment, the frequency deviation is deduced from the interference beat signal between the frequency-doubled idler wave and the signal wave. This beat signal is equal to $\{2\omega + \Delta\omega$ minus $2 \times (\omega - \Delta\omega)\}$ to give $3\Delta\omega$. According to standard practice in cavity stabilization, we give this beat signal an offset frequency Ω_{AOM} of +54 MHz by passing the SRO signal beam through an acousto-optic modulator before mixing with the idler's second harmonic. With this offset it is then possible to track the deviation $\Delta\omega$ on both sides of $\Delta\omega = 0$ readily. The main ICs employed in our phase detection circuit are two ultrafast comparators (Analog Device AD96685), one phase discriminator (Analog Device AD9901) and one differential receiver amplifier (Analog Device AD8130). The error signal from our phase detection circuit is sent into the high-speed proportional integrating controller. Finally, the PZT driver adjusts a PZT (mounted on M3) in accordance to output from the proportional controller.

The quality of the stabilization control is monitored by recording the beat signal thus produced using a radio frequency (RF) spectrum analyzer (Rohde & Schwarz FSL3) with 100 kHz resolution bandwidth. The frequency of the recorded beat signal is $3\Delta\omega + \Omega_{AOM}$. Histograms of the recorded frequency deviation over a 30 min period are shown in Figure 2 with and without cavity stabilization when the temperature of the OPO crystal was set to 92.8 °C to achieve phase-matching in the frequency division-by-three cw SRO. Figure 2a shows the frequency deviation, $\Delta v = \Delta\omega/2\pi$, without cavity stabilization. The recorded deviation $\Delta\omega$ centers at 131.8 MHz with a ±14.7-MHz random drift (one standard deviation). With the stabilization circuit turned on, the frequency deviation is reduced to centering at -76 kHz with a ±437-kHz random drift. The solid red lines in Figure 2a,b are Gaussian fits of the distribution indicating that the source of the drifts are randomly distributed. The drift width in Figure 2b of 437 kHz is comparable to the frequency stability of the seed DFB diode laser of the fiber amplifier, implying that the seed laser's drifts may be determining the width of the stabilized source. This result shows that by locking the seed laser to a stabilized reference-frequency comb linked to a primary frequency standard $\Delta\omega$ can be reduced to less than one Hz and phase-stabilized eventually for waveform synthesis.

126

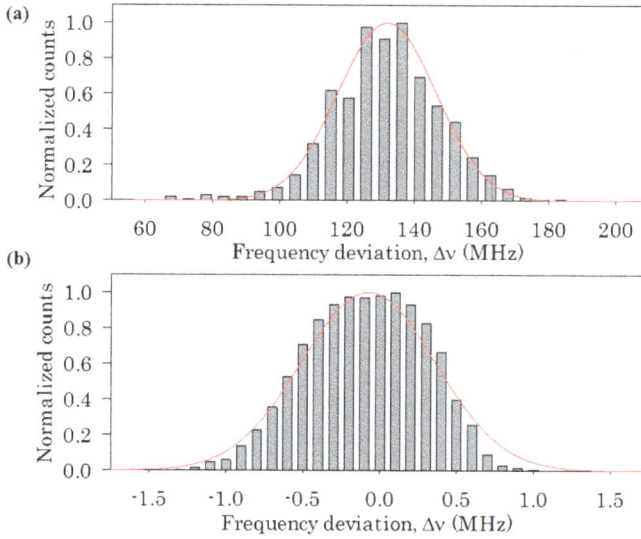

Figure 2. The histograms of frequency deviation from 30-min beat-wave signal recording by a radio frequency (RF) spectrum analyzer. The solid red line is fitting curve based on Gaussian distribution (**a**) without cavity stabilization. The frequency deviation centers at 131.8 MHz with a 14.7-MHz standard deviation; (**b**) with cavity stabilized feedback circuit functioning properly. The center is at −76 kHz with a 437-kHz standard deviation.

3.2. Power of Harmonic Comb Components

For a pump wavelength of 1064 nm, the generated harmonics are at 3192 nm, 1596 nm, 1064 nm, 798 nm, 638.4 nm, and 532 nm. Output power from the SRO and subsequent harmonics produced are measured with a broadband thermopile detector (Coherent P10). The idler power at 3192 nm is determined after its transmission through the SHG crystal and a dichroic mirror used to block out the second harmonic and any residual pump power. The third harmonic (residual pump) is combined with the second, fourth, fifth and sixth harmonics that are collinear after their generation. A Pellin Broca prism is used to disperse these harmonics before sending each into the power meter. The harmonics power is measured as a function of the input pump power. The results are shown in Figure 3.

127

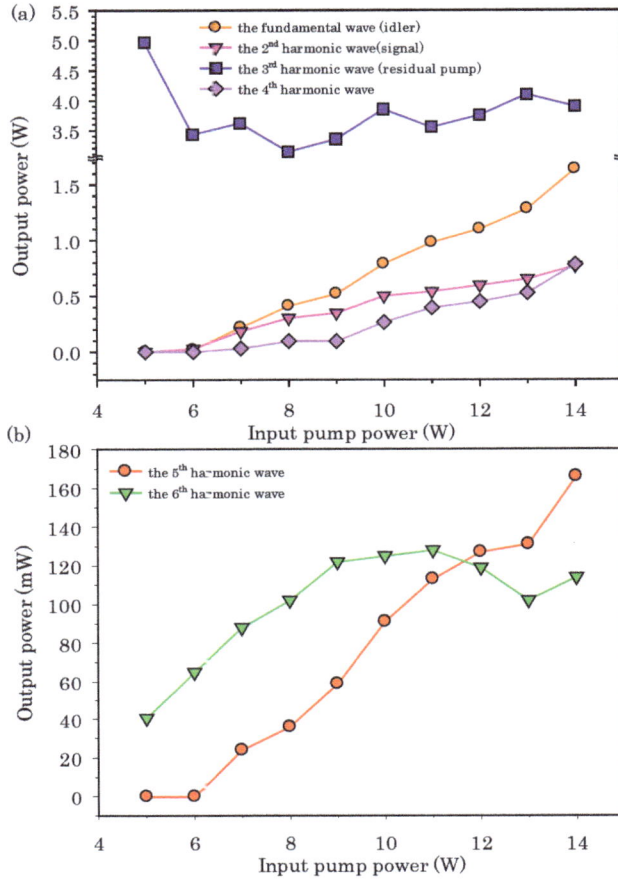

Figure 3. The measured power of the multi-harmonic optical comb after Pellin Broca prism and the dichroic mirror after idler SHG in Figure 1. Here, (**a**) shows the power of the fundamental wave (idler), the 2nd harmonic wave (signal), the 3rd harmonic wave (residual pump power), and the 4th harmonic wave which were generated from the frequency divided-by-three SRO and the intracavity second harmonic generation for the 4th harmonic wave; (**b**) shows the power of 5th harmonic wave and the 6th harmonic wave. The 5th harmonic wave and the 6th harmonic wave are generated from single pass sum frequency generation and second harmonic generation, respectively.

Figure 3a shows the power of the fundamental wave (idler), the 2nd harmonic wave (signal), the 3rd harmonic wave (residual pump power), and the 4th harmonic wave in the multi-harmonic optical comb. The SRO's lasing threshold is at 6 W. This relatively high threshold is because the length of the SRO nonlinear crystal adopted

128

in this experiment is shorter than used in previous experiments, and the insertion loss from the etalon and the intracavity SHG crystal. Without these extra optics, the SRO threshold drops to about 2 W. The maximum fundamental wave (idler), and the 2nd harmonic wave (signal) power are 1.64 W and 0.77 W, respectively at an input pump power of 14 W. The residual pump power (the 3rd harmonic wave) at this input is 3.9 W. For an output coupling of 0.6% the estimated circulating signal power in the cavity is as much as 128 W. This accounts for a respectable cw SHG conversion of the signal to the 4th harmonic which is measured to be 0.78 W without correcting for losses. The 5th harmonic and the 6th harmonic are obtained by single pass wavelength mixing. Figure 3b indicates the maximum powers obtained of the 5th harmonic wave and the 6th harmonic wave are 166 mW and 114 mW, respectively.

The power achieved for the frequency division-by-three SRO is at least ten times higher than previously reported in the literature [17]. This is the first time a cw harmonic comb of up to six harmonics has been reported. The total harmonic comb power of 7.4 W is unprecedented. High cw power is needed for effective phase and amplitude management of the comb components in waveform synthesis [19] and its subsequent applications. The multiwatt comb power that has been achieved is expected to be sufficient for this purpose.

3.3. Simulated Waveform Synthesis with cw Harmonic Comb

The harmonic comb that is reported here has a fundamental wavelength of 3192 nm. The relative phase relationship of the harmonic comb is critical during waveforms synthesis. A fixed phase relationship in phase-matched three wave mixing process is $\phi_a = \phi_b + \phi_c - \pi/2$, where subscripts a, b, and c represent the identities of the three optical waves [20]. Since all components of the cw harmonic comb in this work are generated from three wave mixing process; the deduced phase of ω_n is $\phi_n = n\phi_1 - (n - 1)\pi/2$, where $\phi_1 = (\phi_3 + \pi)/3$. In this harmonic comb, the phase of the pump, ϕ_3 is the only unchangeable parameter in the system. The other components' phase will follow the phase of pump based on the deduced relation. By managing the phase and amplitude of each comb component periodic field waveforms of arbitrary shape could be synthesized. The calculated repetition rate of the synthesized pulse train with this comb is ~94 THz, and has an equivalent period spacing of 10.6 fs. The shortest pulse synthesized within each period will be a transform-limited sub-cycle cosine pulse that has an electric field FWHM of 942 attoseconds. The FWHM of the intensity envelope of this cosine pulse is 1.59 fs. The simulated waveforms, synthesized with the comb produced in this experiment and when the spectral phases are adjusted to be equal, are plotted and shown as the red curve in Figure 4. By adjusting the amplitudes of the comb components to be equal, the narrowest pulses obtainable with this comb are plotted in blue in the figure.

For a comb produced with a perfect divided-by-three cw SRO this pulse train will be continuous. In reality the SRO is not perfect and there is a frequency deviation $\Delta\omega$ as described in Section 3.1 above. The time duration in which the pulses in the train will remain intact is dependent on this frequency deviation $\Delta\omega$ of the SRO. With $\Delta\omega$ at one standard deviation of the present case the period of repetition of the pattern of synthesized pulses is ~2288 ns, corresponding to the time inverse of $\Delta\omega$ of 437 kHz. Figure 5 shows the evolution of the synthesized waveform at this frequency deviation over a time span of 3000 ns, clearly indicating the pattern repeats itself every 2288 ns. Allowing for a drop of 10% in the maximum field strength as the criteria, then according to the numerical simulations, the sub-cycle cosine electric field maintains its strength for over 376 ns in every cycle. Since this is for a one standard deviation of the frequency deviation, 86% of the time this waveform has this shape for 376 ns or longer.

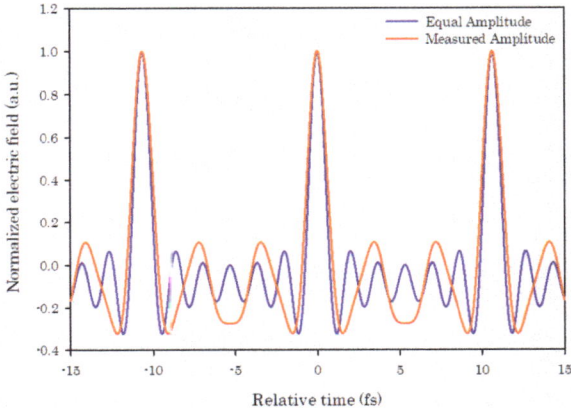

Figure 4. Synthesized waveforms calculated for the six component comb obtained in this experiment (red curve) and for the case where the components are reduced to equal amplitude (blue curve) for the zero detuning case. This will produce a continuous train of pulses of the calculated shape shown here.

We pointed out in Section 3.1 that the residual frequency deviation is due to the drift of the seed laser. Hence we believe that this duration can be extended to over 1000 ns by actively stabilizing the seed laser of the pump in this experiment.

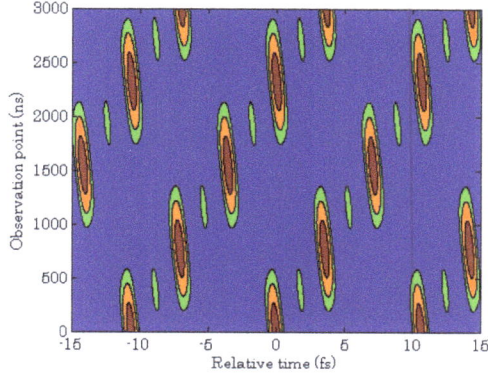

Figure 5. Numerical simulations of a subcycle cosine pulse electric field generated by this multi-harmonic optical comb with a frequency deviation of 437 kHz. The vertical scale is the time evolution of the pulse train from 0 to 3000 ns. The horizontal scale is the local time of the waveform displayed for about three cycles. The field pattern shifts with a phase shift of $2\pi/3$ every 1/3 of a period by repeats and replicates itself in approximately 2288 ns which corresponds to the inverse of the frequency deviation used in the simulation. Normalized electric field strength shown here is color-coded to assist in recognition of the evolution of the field pattern as follows: dark red regions represent the value is from 0.8 to 1.0; orange regions represent the value is from 0.5 to 0.8; green regions represent the value is from 0.253 to 0.5 and blue regions represent the value is from 0.0 to 0.253.

4. Conclusions

We have demonstrated a broadband cw multi-harmonic optical comb based on a frequency divided-by-three optical parametric oscillator. The frequency deviation is centered at −76 kHz with a fluctuation of 437-kHz in one standard deviation. According to our numerical simulation, a stable subcycle cosine electric field pulse can last for more than 376 ns at 437-kHz frequency deviation, which is two orders of magnitude longer than in any synthesizing schemes that have been reported. The output powers of the spectral components in this cw optical comb are 1.64 W, 0.77 W, 3.9 W, 0.78 W, 0.17 W, and 0.11 W which correspond to the fundamental wave to the 6th harmonics wave. The bandwidth of this multi-harmonic comb is ~470 THz. The results show that this cw multi-harmonic optical comb can be a useful light source of a stable optical waveform function generator.

Acknowledgments: We thank Shou-Tai Lin, Chia-Chen Hsu, Jin-Long Peng, Shang-Da Yang, Chen-Bin Huang, and Yuan-Yao Lin for helpful discussions. We acknowledge financial support by the Ministry of Science and Technology of Taiwan, Grants 101-2120-M-007-002, 101-2112-M-001-008, 101-2221-E-007-105 and 102-2221-E-007-111, the Academia Sinica of Taiwan, the Ministry of Education of Taiwan and the National Tsing Hua University.

Author Contributions: Drafting of manuscript: Yen-Yin Lin; Acquisition of data: Po-Shu Wu, Hsiu-Ru Yang; Analysis and interpretation of data: Po-Shu Wu, Hsiu-Ru Yang, Yen-Yin Lin; Critical revision: Yen-Yin Lin, A. H. Kung; Planning and supervision of the research: Yen-Yin Lin, Jow-Tsong Shy, A. H. Kung.

Conflicts of Interest: The authors declare no conflict of interest.

References

1. Hänsch, T.W.; Walther, H. Laser Spectroscopy and Quantum Optics. *Rev. Mod. Phys.* **1999**, *71*, S242.
2. Hänsch, T.W. A Proposed Sub-Femtosecond Pulse Synthesizer Using Separate Phase-Locked Laser Oscillators. *Opt. Commun.* **1990**, *80*, 71–75.
3. Yavuz, D.D. Toward Synthesis of Arbitrary Optical Waveforms. *Science* **2011**, *331*, 1142–1143.
4. Chan, H.-S.; Hsieh, Z.-M.; Liang, W.-H.; Kung, A.H.; Lee, C.-K.; Lai, C.-J.; Pan, R.-P.; Peng, L.-H. Synthesis and Measurement of Ultrafast Waveforms from Five Discrete Optical Harmonics. *Science* **2011**, *331*, 1165–1168.
5. Goulielmakis, E.; Yakovlev, V.S.; Cavalieri, A.L.; Uiberacker, M.; Pervak, V.; Apolonski, A.; Kienberger, R.; Kleineberg, U.; Krausz, F. Attosecond Control and Measurements: Lightwave Electronics. *Science* **2007**, *317*, 769–775.
6. Serebryannikov, E.E.; Goulielmakis, E.; Zheltikov, A.M. Generation of Supercontinuum Compressible to Single-Cycle Pulse Widths in an Ionizing Gas. *New J. Phys.* **2008**, *10*, 093001.
7. Chin, S.L. *Femtosecond Laser Filamentation*; Springer-Verlag: New York, NY, USA, 2009.
8. Harris, S.E.; Sokolov, A.V. Subfemtosecond Pulse Generation by Molecular Modulation. *Phys. Rev. Lett.* **1998**, *81*, 2894–2897.
9. Nazarkin, A.; Korn, G.; Wittman, M.; Elsaesser, T. Group-Velocity-Matched Interactions in Hollow Waveguide: Enhanced High-Order Raman Scattering by Impulsively Excited Molecular Vibrations. *Phys. Rev.* **2002**, *65*, 041802.
10. Kawano, H.; Hirakawa, Y.; Imasaka, T. Generation of More than 40 Rotational Raman Lines by 336 Picosecond and Femtosecond Ti:sapphire Laser for Fourier Synthesis. *Appl. Phys. B Lasers Opt.* **1997**, *65*, 1–4.
11. McCracken, R.A.; Sun, J.; Leburn, C.G.; Reid, D.T. Broadband Phase Coherence between an Ultrafast Laser and an OPO using Lock-to-Zero CEO Stabilization. *Opt. Express* **2012**, *20*, 16269–16274.
12. Huang, S.W.; Cirmi, G.; Moses, J.; Hong, K.H.; Bhardwaj, S.; Birge, J.R.; Chen, L.J.; Li, E.; Eggleton, B.J.; Cerullo, G.; *et al.* High-Energy Pulse Synthesis with Sub-Cycle Waveform Control for Strong-Field Physics. *Nat. Photonics* **2011**, *5*, 475–479.
13. Wirth, A.; Hassan, M.T.; Grguraš, I.; Gagnon, J.; Moulet, A.; Luu, T.T.; Pabst, S.; Santra, R.; Alahmed, Z.A.; Azzeer, A.M.; *et al.* Synthesized light transients. *Science* **2011**, *334*, 195–200.

14. Manzoni, C.; Huang, S.W.; Cirmi, G.; Farinello, P.; Moses, J.; Kärtner, F.X.; Cerullo, G. Coherent synthesis of ultra-broadband optical parametric amplifiers. *Opt. Lett.* **2012**, *37*, 1880–1882.

15. Green, J.T.; Weber, J.J.; Yavuz, D.D. Continuous-Wave Light Modulation at Molecular Frequencies. *Phys. Rev.* **2010**, *82*, 011805.

16. Ulvila, V.; Phillips, C.R.; Halonen, L.; Vainio, M. Frequency Comb Generation by a Continuous-Wave-Pumped Optical Parametric Oscillator Based on Cascading Quadratic Nonlinearities. *Opt. Lett.* **2013**, *38*, 4281–4284.

17. Douillet, A.; Zondy, J.-J.; Santarelli, G.; Makdissi, A.; Clairon, A. A Phase-Locked Frequency Divide-by-3 Optical Parametric Oscillator. *IEEE Tran. Instrum. Meas.* **2001**, *50*, 548–551.

18. Lee, D.-H.; Klein, M.E.; Meyn, J.-P.; Richard, W.R.; Gross, P.; Boller, K.-J. Phase-Coherent All-Optical Frequency Division by Three. *Phys. Rev.* **2003**, *67*, 013808.

19. Hsieh, Z.-M.; Lai, C.-J.; Chan, H.-S.; Wu, S.-Y.; Lee, C.-K.; Chen, W.-J.; Pan, C.-L.; Yee, F.-G.; Kung, A.H. Controlling the Carrier-Envelope Phase of Raman-Generated Periodic Waveforms. *Phys. Rev. Lett.* **2009**, *102*, 213902.

20. Armstrong, J.A.; Bloembergen, N.; Ducuing, J.; Pershan, P.S. Interactions between light waves in a nonlinear dielectric. *Phys. Rev.* **1962**, *127*, 1918–1939.

Chapter 2:
Measurement of Ultrashort Optical Pulse

Autocorrelation and Frequency-Resolved Optical Gating Measurements Based on the Third Harmonic Generation in a Gaseous Medium

Yoshinari Takao, Tomoko Imasaka, Yuichiro Kida and Totaro Imasaka

Abstract: A gas was utilized in producing the third harmonic emission as a nonlinear optical medium for autocorrelation and frequency-resolved optical gating measurements to evaluate the pulse width and chirp of a Ti:sapphire laser. Due to a wide frequency domain available for a gas, this approach has potential for use in measuring the pulse width in the optical (ultraviolet/visible) region beyond one octave and thus for measuring an optical pulse width less than 1 fs.

Reprinted from *Appl. Sci.* Cite as: Takao, Y.; Imasaka, T.; Kida, Y.; Imasaka, T. Autocorrelation and Frequency-Resolved Optical Gating Measurements Based on the Third Harmonic Generation in a Gaseous Medium. *Appl. Sci.* **2015**, *5*, 136–144.

1. Introduction

Many papers report attosecond pulse generation via high-order harmonic generation (HHG) for use in the studies of ultrafast phenomena related to electrons in an inner-shell orbital of an atom and molecule [1]. On the other hand, an ultrashort optical pulse approaching 1 fs, the spectrum of which is extended from the ultraviolet (UV) to the visible (VIS) region, would be useful for studies of ultrafast phenomena related to electrons in an outer-shell orbital, which are directly concerned with chemical bond/reaction and are more important in chemistry. For example, a molecular ion in mass spectrometry can be strongly enhanced by decreasing the optical pulse width in the femtosecond regime, which is then useful for more reliable identification of the explosive substances [2]. Several methods have been developed for generating an extremely-short optical pulse. For example, numerous emission lines have been generated in the entire VIS region, based on non-resonant four-wave mixing in a thin fused silica plate. This technique has been utilized to generate 2.2-fs optical pulses [3]. On the other hand, coherent supercontinua have been generated from the UV to near-infrared (NIR) by focusing the beam in a hollow-core fiber filled with neon gas, and a three-channel optical field synthesizer has been utilized to generate 2.1-fs pulse [1]. As recognized from the uncertainty principle, a wider spectral domain is essential for the generation of a 1-fs optical pulse. The generation of high-order Raman sidebands extending from the deep-UV (DUV) to the NIR region was first reported in a few decays ago [4]. To date, an extremely wide spectral region extending from 183 to 1203 nm has been covered, based on resonant vibrational four-wave Raman mixing in molecular hydrogen, suggesting that the generation of a 1-fs optical pulse by a phase control of the emission lines is possible [5].

A variety of techniques, including autocorrelation (AC) and frequency-resolved optical gating (FROG), have been developed to measure optical pulse widths [6]. In these techniques, a nonlinear optical effect such as second harmonic generation (SHG), self-diffraction (SD), and others have been utilized. To measure a 1-fs optical pulse, it is necessary to use a nonlinear optical effect that is usable in a wide frequency domain in the UV-VIS region. A GaN diode with a nature of two-photon absorption in the VIS region (at round 400 nm) has been employed as a detector in a fringe-resolved autocorrelator (FRAC) [7]. In a previous study, we reported on the development of an FRAC, in which a mass spectrometer was employed as a two-photon-response device in the DUV region (at around 267 nm) [8]. However, the frequency domain of these techniques is limited to one octave [9]. As a result, it is difficult to measure optical pulse widths less than 1 fs in the UV-VIS region. In order to overcome this problem, the use of a nonlinear optical effect that can be used in the spectral region wider than one octave would be necessary. One of the approaches would be the use of a third harmonic generation (THG) as a nonlinear optical effect, since it has a frequency domain of twice one octave [10]. A surface-sensitive THG has been successfully used for the autocorrelation measurement [11,12]. Moreover, several papers have reported on the FROG system based on THG (THG-FROG), including the THG on the surface of a glass plate [13], in organic films [14], and in a glass coverslip used in multiphoton microscopy [15]. The use of a solid material is simple and easy-to-use. It is, however, difficult to transmit the THG beam generated in the vacuum-UV (VUV) region and to reduce dispersion to negligible levels, especially in the DUV region. On the other hand, a gas such as helium or argon is transparent, even in the VUV region, and the dispersion sufficiently small to be negligible. Thus, the THG in a gaseous medium would be useful for measuring a pulse width in a wide frequency domain. To our knowledge, a THG-based technique such as THG-FROG using a gaseous medium has not been reported to date.

In this study, we used argon or air as a nonlinear optical medium for THG to measure the pulse width of a fundamental beam of a Ti:sapphire laser emitting at 800 nm as a proof-of-principle experiment. Two types of AC systems, *i.e.*, fringe-resolved AC (FRAC) and intensity-AC (IAC), were developed and were utilized for measuring the pulse width. In addition, a FROG system was developed that permits the pulse width and the chirp of the pulse to be evaluated more accurately. The results were compared with values obtained using FRAC and IAC.

2. Experimental Section

2.1. Fringe-Resolved Autocorrelation

Figure 1 shows a block diagram of the instrument used in this study. In measuring an AC trace, a fundamental beam of a Ti:sapphire laser (Elite, 800 nm, 35 fs, 1 kHz, 4 mJ, Coherent) was passed through a Mach-Zehnder interferometer, consisting of beam splitters made of BK7 plates with a thickness of 3 mm. Using a lens (BK7) with a focal length of 300 mm, the aligned beam was focused into argon gas contained in a cell (700 mm long, 10 mm i.d.) equipped with fused silica windows with a thickness of 0.5 mm. The third harmonic emission generated in the optical components such as the beam splitters was suppressed by passing the beam through a cover glass for microscopy and was confirmed to be negligible levels by evacuating the gas cell. The third harmonic emission generated in the argon gas was passed through an aqueous solution of $NiSO_4$ (500 g/L in water) to suppress the fundamental beam; a substrate of Si was used to remove the fundamental beam of a Ti:sapphire laser in the reported work [16]. The THG beam emitting at 267 nm was further isolated using a monochromator and was measured using a photomultiplier (R1332, no response at 800 nm, Hamamatsu Photonics) designed for measuring a large optical pulse. In the experiment, the signal intensity was measured at the level sufficiently lower than the level of signal saturation. The output signal was fed into a boxcar integrator, and the averaged signal was recorded using an oscilloscope. A function generator provided a signal for a piezoelectric transducer equipped with a translational stage with a retroreflector mounted on it. An autocorrelation trace was collected using the oscilloscope by recording the signal against the time delay between the pulses.

2.2. Frequency-Resolved Optical Gating

A FROG trace was measured using the instrument shown in Figure 2. The fundamental beam of the Ti:sapphire laser was separated into two parts using a pair of D-shape mirrors to remove the dispersion arising from the beam splitters. The beams were reflected by means of a pair of aluminum mirrors and were focused with an off-axis parabolic mirror into a nonlinear optical medium such as argon or air. The beam was collimated using a fused silica lens with a focal length of 10 cm and was introduced into a slit of a monochromator (CT-10, Jasco) using a rotating mirror. One of the THG beams, corresponding to $E_{sig}(t, \tau) = E(t)^2 E(t - \tau)$, was measured as a signal. Electronics similar to those used in the FRAC experiment were employed for measuring the third harmonic emission. The spectrum of the THG beam was measured using the second-order diffraction of the grating to improve the spectral resolution of the monochromator. A FROG trace was measured by scanning the wavelength of the monochromator at different positions of the delay for one of the

D-shape mirrors. The data were measured and analyzed using a program provided by Femtosoft Technologies.

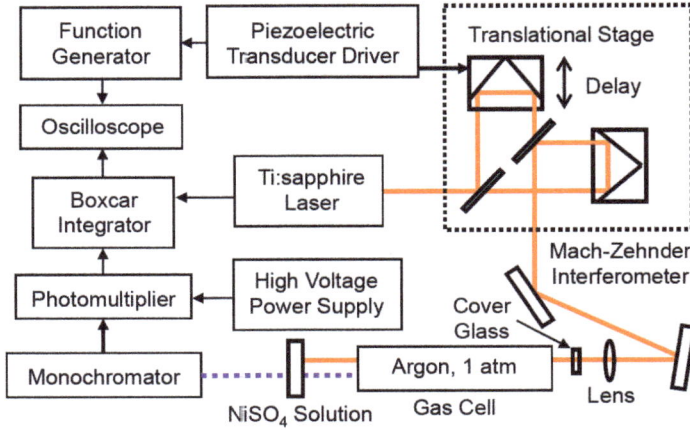

Figure 1. Experimental apparatus for third harmonic generation-fringe-resolved autocorrelator (THG-FRAC). The orange (solid) and blue (broken) lines show the fundamental and THG beams, respectively.

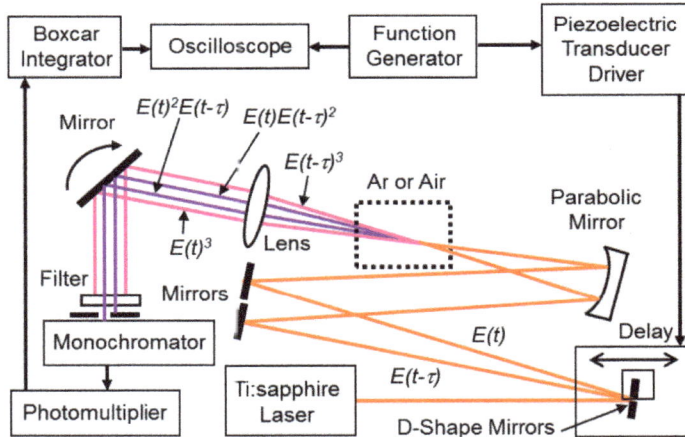

Figure 2. Experimental apparatus for THG-frequency-resolved optical gating (FROG).

2.3. Intensity Autocorrelation

An IAC trace was obtained using a non-collinear configuration developed for use in a FROG system shown in Figure 2. Thus, the instrument consists of only reflective optics except for a window of the gas cell, allowing a nearly dispersion-free experiment.

140

3. Results and Discussion

3.1. Fringe-Resolved Autocorrelation

The spectral width measured for the fundamental beam of the Ti:sapphire laser was 27 nm (420 cm^{-1}). The Fourier-transform-limited (FTL) pulse width calculated by assuming a Gaussian temporal profile was 35 fs, identical to the value provided by the manufacturer of the laser. An FRAC trace is shown in Figure 3A. Suppression of the signal to the background level by destructive interference, in addition to a full modulation of the signal shown in the expanded view (Figure 3B), suggests the pump beams were superimposed in generating the THG beam. The ratio of the signal and the background was *ca.* 25, slightly smaller than the predicted value of 32 for an FTL pulse using the third-order nonlinear effect such as THG [10]. Figure 3C shows the FRAC trace calculated for FTL pulse. The observed trace is slightly wider than this trace, suggesting the chirp of the pulse. Figure 3D shows the FRAC trace calculated under the assumption that the FTL pulse (35 fs) is chirped to 41 fs with a group delay dispersion (GDD) of 500 fs^2, which is nearly identical to the value of GDD (*ca.* 400 fs^2) roughly estimated from the thickness of the optical components in the beam path. The pulse width calculated from the IAC trace, which can be obtained by low-pass filtering the data shown in Figure 3A, was 41 fs, which is in good agreement with the above value. Another possible explanation for the discrepancy between the results of (A) and (C) would be the error in the measurement of the spectrum since a wider band width would be observed due to a finite resolution of the spectrometer, which provides a shorter transform-limited pulse width; see a lack of small wings at round ±50 fs in the observed data (A), which is in contrast to the calculated data (D), suggesting that the effect of chirp is minimal.

3.2. Intensity Autocorrelation

The IAC trace observed in this study is shown in Figure 4. The full width at half maximum (FWHM) of the trace was 44 fs, suggesting a pulse width of 36 fs [10]. This value is slightly smaller than that obtained using an FRAC system, which can be attributed to the use of the reflective optics with no dispersion in the IAC system.

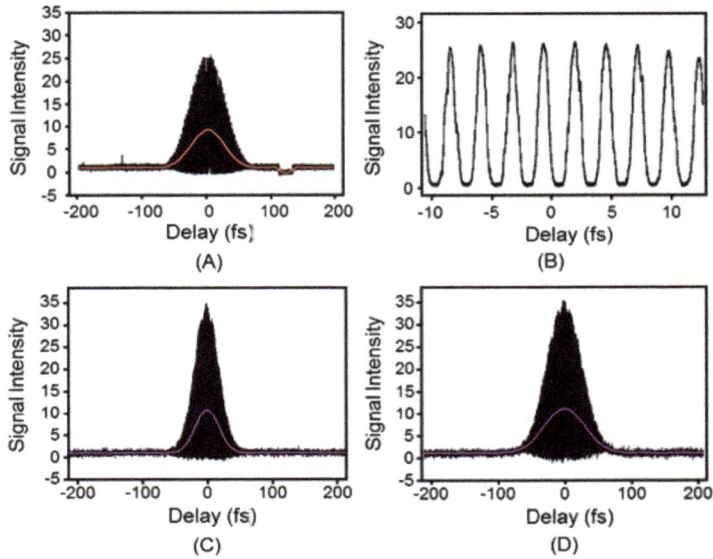

Figure 3. Autocorrelation trace for (**A**) observed data (**B**) expanded view (**C**) theoretically predicted trace for a Fourier-transform-limited (FTL) pulse (**D**) theoretically predicted trace for a chirped pulse (GDD = 500 fs^2). A random noise was calculated using a computer and was added to the calculated data for better visual comparison. In order to check the baseline level of the observed trace (A), the laser beam was interrupted during a period of 110–130 fs in the experiment, suggesting that the signal was suppressed to zero at the bottom of the trace.

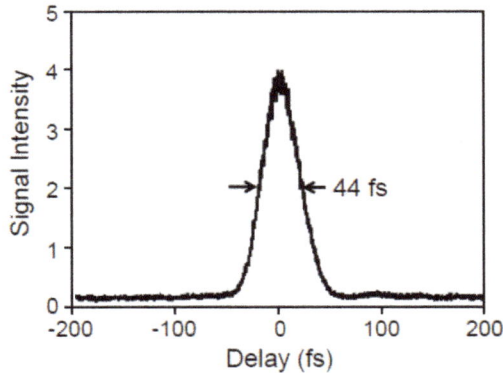

Figure 4. Intensity autocorrelation trace observed for the Ti:sapphire laser.

3.3. Frequency-Resolved Optical Gating

The chirp of the pulse can be more accurately evaluated using a FROG system. The efficiency of THG induced in air was similar to that obtained using argon. Because of this, air was used as a nonlinear optical medium in this study for the sake of simplicity. Before recording the FROG trace, the THG beam pattern was measured to properly extract the signal of $E_{sig}(t, \tau) = E(t)^2 E(t - \tau)$ [6]. The THG beams corresponding to $E(t)^3$, $E(t)^2 E(t - \tau)$, $E(t) E^2(t - \tau)$, and $E^3(t - \tau)$ could be clearly observed by rotating the angle of the mirror (see the arrow in Figure 2). Therefore, the intensity of the THG beam for $E_{sig}(t, \tau) = E(t)^2 E(t - \tau)$ was monitored for measuring the FROG trace. The observed and retrieved data obtained for negatively-chirped, FTL, and positively-chirped pulses are shown in Figure 5. The pulse width measured for an FTL pulse was 32 fs, which appears to be similar to or slightly shorter than the values obtained using IAC. On the other hand, the chirped pulse provided a slightly longer pulse width (positive 42 fs, negative 44 fs) due to the GDD value being calculated to be 300 fs^2.

The minimal value of the pulse width measured using conventional SHG-FROG or SD-FROG is determined by the spectral region that is usable as a nonlinear optical medium: although a broad bandwidth spanning a multi-octave frequency domain can be covered using a thin optical crystal, the spectral region is practically limited to the VIS-NIR region [17]. On the other hand, SHG, which would restrict the spectral region of THG, does not occur in an isotropic gas, and a noble gas such as argon is transparent and has a small dispersion in the VUV to IR region. As a result, the present approach using a gas in conjunction with THG-FROG can be utilized to measure pulse widths in a wide frequency domain extending twice one octave. Because of this, this technique would be applied to the measurement of an ultrashort optical pulse less than 1 fs especially in the VUV-DUV region: it should be noted that a higher carrier frequency is desirable for generating a shorter optical pulse. However, in order to avoid the dispersion arising from the optics such as cell windows, it would be necessary to use a nozzle for introduction of a rare gas (not air) into a vacuum. Efficient generation of THG (even HHG) reported to date suggests sufficient sensitivity of this THG-based method using a gas for the measurement of an ultrashort pulse width.

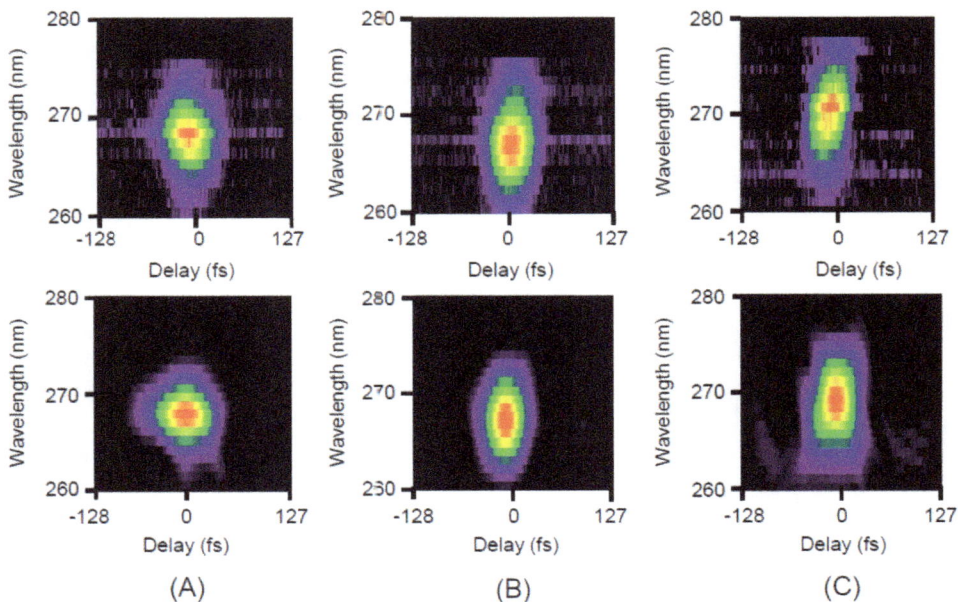

Figure 5. THG-FROG traces obtained for (**A**) negatively-chirped (**B**) FTL (**C**) positively-chirped pulses. The chirp of the pulse was adjusted by changing the position of the grating in the compressor of the Ti:sapphire laser to the value, at which the energy of the THG pulse decreased to a half of the value obtained using the FTL pulse. Above, original traces; below, retrieved data. FROG error: (**A**) 0.6% (**B**) 0.6% (**C**) 1%.

4. Conclusions

In this study, we report on the development of FRAC, IAC, and FROG systems based on THG using a gaseous medium. This approach can be used to measure an ultrashort optical pulse at any wavelength from the UV to the IR region although the emission generated by THG is located in the VUV-VIS region. Therefore, this technique can be applied to the lasers used in various areas of science and technology.

Acknowledgments: This research was supported by a Grant-in-Aid for the Global COE program, "Science for Future Molecular Systems" from the Ministry of Education, Culture, Sports, Science and Technology of Japan and by Grants-in-Aid for Scientific Research from the Japan Society for the Promotion of Science (JSPS) KAKENHI Grant Number 23245017, 24510227, 26220806, and 15K13726. This study also received support from the Steel Industry Foundation for the Advancement of Environmental Protection Technology.

Author Contributions: Performing the experiment: Yoshinari Takao, Numerical simulation: Tomoko Imasaka, Drafting of manuscript: Totaro Imasaka, Critical revision: Yuichiro Kida, Planning and supervision of the research: Totaro Imasaka.

Conflicts of Interest: The authors declare no conflict of interest.

References

1. Wirth, A.; Hassan, M.Th.; Grguraš, I.; Gagnon, J.; Moulet, A.; Luu, T.T.; Pabst, S.; Santra, R.; Alahmed, Z.A.; Azzeer, A.M.; *et al.* Synthesized light transients. *Science* **2011**, *334*, 195–200.

2. Hamachi, A.; Okuno, T.; Imasaka, T.; Kida, Y.; Imasaka, T. Resonant and nonresonant multiphoton ionization processes in the mass spectrometry of explosives. *Anal. Chem.* **2015**, *87*, 3027–3031.

3. Weigand, R.; Mendonça, J.T.; Crespo, H.M. Cascaded nondegenerate four-wave-mixing technique for high-power single-cycle pulse synthesis in the visible and ultraviolet ranges. *Phys. Rev. A* **2009**, *79*, 063838.

4. Imasaka, T.; Kawasaki, S.; Ishibashi, N. Generation of more than 40 laser emission lines from the ultraviolet to the visible regions by two-color stimulated Raman effect. *App. Phys. B* **1989**, *49*, 389–392.

5. Shitamichi, O.; Imasaka, T. High-order Raman sidebands generated from the near-infrared to ultraviolet region by four-wave Raman mixing of hydrogen using an ultrashort two-color pump beam. *Opt. Express* **2012**, *20*, 27959.

6. Trebino, R. *Frequency-Resolved Optical Gating: The Measurement of Ultrashort Laser Pulses*; Kluwer Academic Publishers: Boston, MA, USA, 2002.

7. Zürch, M.; Hoffmann, A.; Gräfe, M.; Landgraf, B.; Riediger, M.; Spielmann, Ch. Characterization of a broadband interferometric autocorrelator for visible light with ultrashort blue laser pulses. *Opt. Commun.* **2014**, *321*, 28–31.

8. Zaitsu, S.; Miyoshi, Y.; Kira, F.; Yamaguchi, S.; Uchimura, T.; Imasaka, T. Interferometric characterization of ultrashort deep ultraviolet pulses using a multiphoton ionization mass spectrometer. *Opt. Lett.* **2007**, *32*, 1716–1718.

9. Imasaka, T.; Imasaka, T. Searching for a molecule with a wide frequency domain for non-resonant two-photon ionization to measure the ultrashort optical pulse width. *Opt. Commun.* **2012**, *285*, 3514–3518.

10. Meshulach, D.; Barad, Y.; Silberberg, Y. Measurement of ultrashort optical pulses by third-harmonic generation. *J. Opt. Soc. Am. B* **1997**, *14*, 2122–2125.

11. Tsang, T.Y.F. Optical third-harmonic generation at interfaces. *Phys. Rev. A* **1995**, *52*, 4116–4125.

12. Squier, J.A.; Fittinghoff, D.N.; Barty, C.P.J.; Wilson, K.R.; Müller, M.; Brakenhoff, G.J. Characterization of femtosecond pulses focused with high numerical aperture optics using interferometric surface-third-harmonic generation. *Opt. Commun.* **1998**, *147*, 153–156.

13. Tsang, T.; Krumbügel, M.A.; Delong, K.W.; Fittinghoff, D.N.; Trebino, R. Frequency-resolved optical-gating measurements of ultrashort pulses using surface third-harmonic generation. *Opt. Lett.* **1996**, *21*, 1381–1383.

14. Ramos-Ortiz, G.; Cha, M.; Thayumanavan, S.; Mendez, J.; Marder, S.R.; Kippenlen, B. Ultrafast-pulse diagnostic using third-order frequency-resolved optical gating in organic films. *Appl. Phys. Lett.* **2004**, *85*, 3348–3350.

15. Chadwick, R.; Spahr, E.; Squier, J.A.; Durfee, C.G.; Walker, B.C.; Fittinghoff, D.N. Fringe-free, background-free, collinear third-harmonic generation frequency-resolved optical gating measurements for multiphoton microscopy. *Opt. Lett.* **2006**, *31*, 3366–3368.
16. Graf, U.; Fieß, M.; Schultze, M.; Kienberger, R.; Krausz, F.; Goulielmakis, E. Intense few-cycle light pulses in the deep ultraviolet. *Opt. Express* **2008**, *16*, 18956.
17. Birkholz, S.; Steinmeyer, G.; Koke, S.; Gerth, D.; Bürger, S.; Hofmann, B. Phase retrieval via regularization in self-diffraction based spectral interferometry. **2014**. arXiv:1412.2965.

Multicolored Femtosecond Pulse Synthesis Using Coherent Raman Sidebands in a Reflection Scheme

Kai Wang, Alexandra A. Zhdanova, Miaochan Zhi, Xia Hua and
Alexei V. Sokolov

Abstract: Broadband coherent Raman generation emerges as a successful method
to produce multicolored femtosecond pulses and time-shaped laser fields. In our
study, coherent Raman sidebands are generated in a Raman-active crystal, driven
by two-color femtosecond laser pulses. An interferogram of the sidebands based on
coherent Raman scattering is produced in a novel reflection scheme. The relative
spectral phases of the sidebands are obtained from the interferogram using a
numerical simulation. This enables us to retrieve the ultrafast waveform using
coherent Raman sidebands.

Reprinted from *Appl. Sci.* Cite as: Wang, K.; Zhdanova, A.A.; Zhi, M.; Hua, X.;
Sokolov, A.V. Multicolored Femtosecond Pulse Synthesis Using Coherent Raman
Sidebands in a Reflection Scheme. *Appl. Sci.* **2015**, *5*, 145–156.

1. Introduction

The generation of subfemtosecond pulses would extend the horizon of ultrafast
measurements to the time scale of electronic motion. Remarkable progress has
been made toward generation and characterization of ever-shorter pulses in the
short-wavelength spectral region. For example, a single isolated attosecond (as) pulse
of 67 as was composed from an extreme UV supercontinuum covering 55–130 eV
generated by the double optical gating technique [1].

On the other hand, a few-femtosecond pulse in the optical region would have
a great deal of potential in the research of ultrafast science and technology [2,3].
There have been several techniques developed to achieve sub-cycle ultrashort pulses.
For example, a subcycle field transient, which spans the infrared, visible, and
ultraviolet spectral regions, was produced with a 1.5-octave three-channel optical
field synthesizer by Wirth *et al.* [4]. The 2.4 fs transient was focused into a krypton-gas
cell to trigger sub-femtosecond electron motion. In another approach, based on
cascaded four wave mixing (CFWM), Imasaka's group reported a high energy
multicolored femtosecond pulse generated in a hydrogen-filled gas cell and hollow
fiber [5]. Moreover, Kung's group reported a multi-watt broadband continuous-wave
multi-harmonic optical comb based on a frequency division-by-three singly-resonant
optical parametric oscillator [6].

Another light source that delivers sub-fs pulses with a spectrum centered on the visible region has been developed and ultimately may lead to optical arbitrary waveform generation (OAWG) [7]. The light source is based on what has been called "molecular modulation" [8,9], which shares visual similarities with the approach for the generation of ultrashort optical pulses proposed by Yoshikawa and Imasaka [10]. It has been predicted that coherent molecular oscillations can produce laser frequency modulation (FM), with a total bandwidth extending over the infrared, visible, and ultraviolet spectral regions, and with the possibility of sub-femtosecond pulse compression [8]. The technique utilizes ideas of electromagnetically induced transparency (EIT) and relies on adiabatic preparation of maximal molecular coherence. The coherence is established by driving the molecular transition with two single-mode laser fields, slightly detuned from the Raman resonance so as to excite a single molecular eigenstate. Molecular oscillation, in turn, modulates the driving laser frequencies, causing the collinear generation of a very broad FM-like spectrum. This broadband light is inherently coherent and allows for sub-femtosecond (attosecond) pulse compression in the visible-UV range. Although attosecond pulses with wavelengths in the extreme ultraviolet and soft-X-ray pulses have been obtained by high-harmonic generation (HHG), the molecular modulation technique has the potential for generating high-energy sub-femtosecond pulses in the soft UV range, which will enable new experiments that exploit electronic resonances in molecules.

In recent years, several groups have made substantial advances in molecular modulation. For example, Kung's group achieved absolute phase control of five discrete optical harmonics (two pump beams and three generated Raman sidebands from H_2 gas), and thus demonstrated the synthesis and measurement of ultrafast waveforms such as square and saw-tooth fields [7]. Katsuragawa's group reported the carrier envelope offset control of octave-spanning Raman comb by using dual-frequency laser radiation locked on a single laser cavity and, simultaneously, its second harmonic [11]. Broadband spectra based on multifrequency Raman generation were also studied in photonic crystal fiber [12] and hollow fibers filled with SF6 [13]. Meanwhile, the molecular modulation method has been extended to the continuous-wave (CW) domain. For example, Yavuz's group has studied CW—stimulated Raman scattering (SRS) inside a high-finesse cavity and demonstrated a continuous-wave optical modulator at 90 THz [14]. Generation of a phase-locked Raman frequency comb has been demonstrated recently in a simple setup consisting of a microchip laser as pump source and two hydrogen-filled hollow-core photonic crystal fibers [15].

Molecular modulation in gas produces high repetition rate, low-energy pulse trains [16]. However, high energy, isolated pulses are generally more useful in studies of ultrafast phenomena. To this end, we have extended the molecular modulation technique to a qualitatively different time regime and to a different state of medium—Raman-active crystals driven by femtosecond pulses. We have demonstrated the mutual coherence of the spectral sidebands generated through molecular modulation in diamond and shown the capability to control spectral phases in a precise and stable manner in a setup that combined manual course adjustment of individual sideband phases with programmable pulse shaping and fine phase (and amplitude) tuning across the full spectrum [17,18]. Recently, we studied coherent transfer of optical orbital angular momentum in multi-order Raman sideband generation [19,20]. The energy of the ultrafast waveform produced in the scheme of References [17,18] is limited by the damage threshold of the pulse shaper. In order to obtain high energy ultrafast waveforms, we design a reflection scheme using spherical mirrors to combine the Raman sidebands [21]. The sidebands and the driving pulses are refocused back to the Raman crystal and the relative spectral phases are retrieved from an interferogram based on nonlinear Raman interaction. Furthermore, using a deformable mirror (DM) to adjust the spectral phases, we demonstrate that the setup is capable of synthesizing ultrafast waveforms using coherent Raman sidebands.

In this paper, we review and expand on our recent work [21]. We start by describing the algorithm used to retrieve the spectral phases from a two-sideband-interferogram, giving the details of our theoretical simulation of Reference [21]. The method is then extended into the scenario of the interferogram composed of more sidebands; we show the standard deviation of the simulation *vs.* original interferogram.

2. Experimental Setup

The experimental setup is shown in Figure 1. We used a Ti:Sapphire amplifier, which outputs 1 mJ per 40 fs pulse (at 1 kHz repetition rate) with central wavelength at 806 nm. Using a beamsplitter (60:40 (R:T)), the pulse was divided into two parts. Forty percent of the beam was used as the pump beam for Raman generation while the other part was used to pump an optical parametric amplifier (OPA). The second harmonic (900 nm) of the idler beam from the OPA was used as the Stokes beam (we follow the coherent anti-Stokes Raman scattering convention and denote the shorter 806 nm wavelength beam as pump and the longer wavelength 900 nm as the Stokes beam). The pulses were vertically polarized. The power of the pump was around 10 mW and the power of the Stokes was around 2 mW. The intensity of the pump at the focus was around 2×10^{11} W/cm^2 [21]. Coherent Raman sidebands were generated when the two beams were crossed on the crystal (1 mm thick synthetic single-crystal diamond) at 3.7 degrees. The angle was theoretically calculated under the standard phase matching conditions considering the higher order Raman generation. It was further optimized experimentally using the generation of the spectrum and spatial beam profile of the sidebands. Further details on the properties of sideband generation in diamond can be found to in Reference [22]. The robustness of the sidebands generation in a Raman active crystal in our experimental setup has also been studied in Reference [23]. In Figure 1, the top picture shows the Raman sidebands generated from our Raman active crystal (the sidebands were generated in PbWO$_4$).

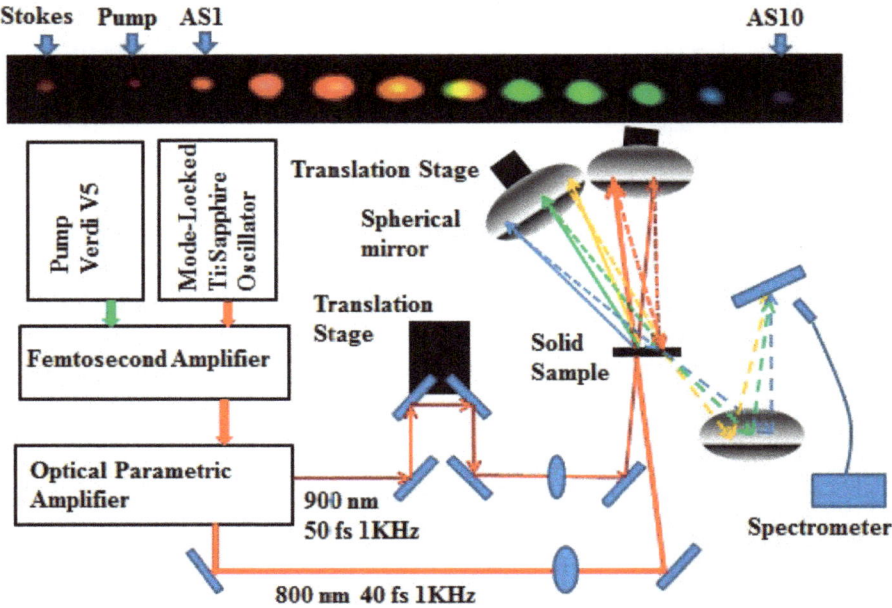

Figure 1. Experimental setup schematic. The Raman sidebands are generated from a Raman active crystal. We used a spherical mirror to reflect the sidebands and the pump and Stokes beams back to the crystal. The top is a picture of the Raman sidebands taken by a camera.

We designed a 2f-2f reflection scheme to characterize the relative spectral phases of the Raman sidebands. Specifically, we used concave spherical mirrors to reflect the beams back to the same crystal. In our experiment, we chose two spherical mirrors with a focal length of 10 cm and kept them about 20 cm away from the crystal. One mirror was to reflect the pump and Stokes beams and the other was to reflect the higher order Raman sidebands (in this paper, AS3, AS4, AS5, AS6, and AS7 are reflected back to the crystal). In this configuration, the beams were re-focused back to the same spot of the crystal and phase matching was automatically fulfilled. Consequently, the interaction between the beams was maximized. The beams were re-focused back to the crystal with an offset from their incident spots such that two

150

spots were just distinguishable from each other. A translation stage was used to finely adjust the distance between the spherical mirror and the crystal. In the experiment, the mirror used to reflect the sidebands was put on the translation stage. The position of the mirror that reflects the pump and Stokes beams was fixed. The configuration enabled us to adjust the relative spectral phases between the pump and Stokes beams and the sidebands. As a result, an interferogram was produced as a function of the relative spectral phases due to the coherent Raman interaction [21]. It was recorded with a 200 nm scanning step of translation stage, which corresponded to a 1.3 fs time delay between the two spherical mirrors. When presenting the experimental data, we converted the scanning of the translation stage to the relative time delay.

3. Results and Discussion

In the reflection scheme, by recording the spectrum as a function of time delay, we obtained an interferogram based on coherent Raman scattering. In this paper, we describe our procedure to retrieve the spectral phases from the inteferogram by using a theoretical simulation.

The experimental details were presented in Reference [21] and all the data presented in this paper is generated from the same experimental setup. The spectra of Raman sidebands were recorded by spectrometer (HR 4000, Ocean Optics). The pulse durations of pump and Stokes were around 50 fs. The pulse durations of the sidebands were a little longer than 50 fs due to the dispersion of the crystal [17,18]. The experimental interferograms are shown in Figure 2a (The range for the spectrometer is from 200 nm–1100 nm and in Figure 2a, we only show the range from 500 nm–700 nm). The interferogram is recorded with an integration time (100 ms) of the spectrometer. Therefore, the spectrogram did not display the shot-to-shot robustness of the sideband generation. The spectrogram shows intensity oscillations due to Raman scattering. A cross-section of the spectrogram at 562.57 nm, which is the central wavelength of AS4, is displayed in Figure 2b. The frequency of the intensity oscillations is roughly equal to the differences between the central frequencies of two sidebands. The peak intensity of the pump and the Stokes beam at the focal point were about 4×10^{11} W/cm^2. These beams will induce the cross phase modulation for the sidebands. Theoretically, the phase shift at 562.57 nm is 7.99 (it is evaluated using the parameter from Reference [24]). During the scanning, when the pump and Stokes beams encountered the sidebands, at the first moment, the intensity of the driving beams increased. The changing of the refractive index is proportional to the intensity, and the spectral phase will change faster when the beam approaches the maximum intensity. This will increase the intensity oscillation frequency on the interferogram. Moreover, when the intensity is decreasing, the intensity oscillation frequency will decrease. In our experiments, we have tried our best to limit the cross phase modulation. By analyzing the spectrogram, we did not see significant variation of the intensity oscillation frequency. However, we could still see that the central frequency of the sidebands is slightly shifted due to the cross-phase modulation induced by pump and Stokes. Another nonlinear effect which plays an important role is self-phase modulation. The high intensity of the pump and the Stokes beams will induce self-phase modulation. It would distort their spectral phases and also broaden their spectrum. All of these would affect the energy and spectrum of the Raman sideband.

The goal of our simulation is to retrieve the relative spectral phases of the sidebands from the interferogram. We emphasize that, in the experiment, the spectrogram was recorded with an integration time 100 ms, which resulted in averaging out the fluctuation of the phase and the energy of the sideband generation (according to Reference [23], the deviation of the sideband's energy is around 15% due to the shot-to-shot fluctuation). Our method is to retrieve the spectral phase for the particular spectrogram and so far we do not consider the shot-to-shot fluctuation in the simulation. Therefore, when judging our retrieval algorithm, the root-mean-square (rms) deviation of the retrieved *vs.* original spectrogram is important. In the experiment, the spectral phase distortion was a combined effect resulting from the dispersion of the crystal and nonlinearities, such as cross phase modulation [17,18,21]. The spherical aberration of the spherical mirror also contributed to the phase distortion. To find the best value for the spectral phases, we compared the theoretical simulation with the experimental interferogram. Therefore, the experimental data was processed for the convenience of comparison with the theoretical simulation. At first, to remove the experimental noise due to laser fluctuations, the data was smoothed. Next, the interferogram was recorded by the spectrometer with a resolution of 0.75 nm; the sampling points were not equally distributed in the frequency domain. Thus, we used 2D bilinear interpolation to reconstruct an interferogram from the experimental data. The reconstructed spectrogram had data points uniformly sampled in the frequency domain with spectral range cut down to 400 nm–700 nm, which covered the spectrum of the sidebands used in our simulation. Between 400 nm and 700 nm, there were 250 sampling points. The reconstructed interferogram is shown in Figure 2c. The vertical axis in the interferogram gives wavlenegth in nanometers, which is converted from ω following $\lambda = 2\pi c / \omega$, here, c is the speed of light. The step of interpolations is 1 fs in the other dimension of the interferogram (which is the time delay between the two spherical mirrors).

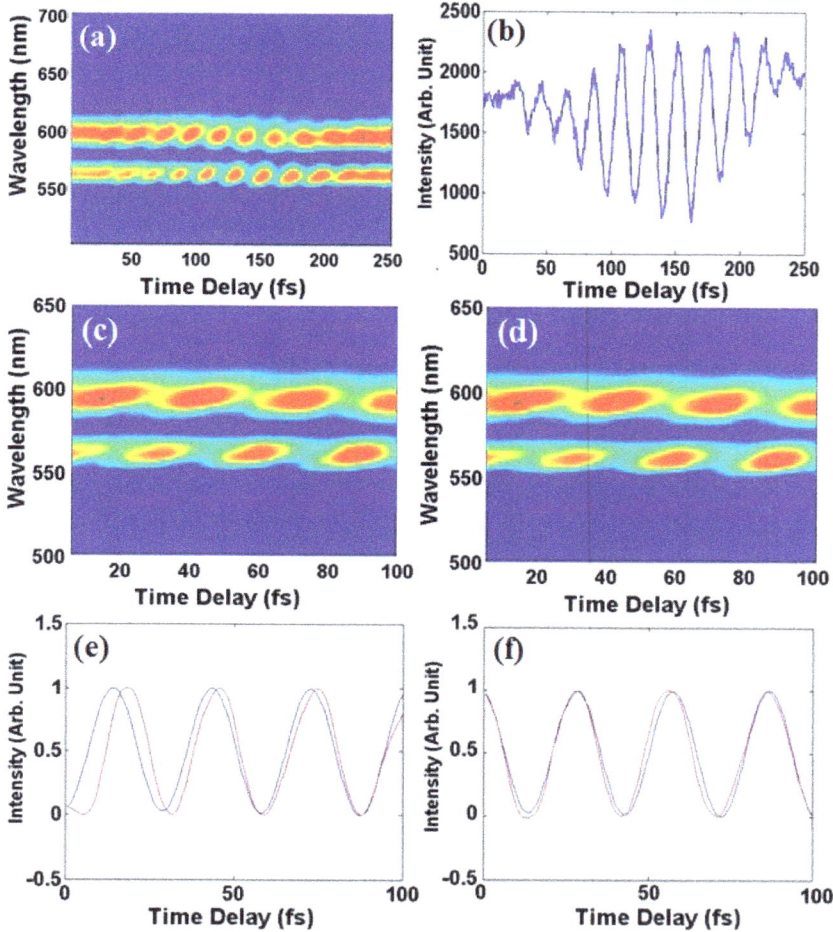

Figure 2. (a) Interferogram of sidebands AS3 and AS4 recorded by spectrometer; (b) The cross-section of the interferogram at 562.57 nm, which is the central wavelength of AS 4, shows the intensity oscillation; (c) Reconstructed interferogram of AS3 and AS4 with the 2D bilinear interpolation; (d) Theoretical simulation of the interferogram using AS3 and AS4; (e) Cross-section of the interferogram from simulation (blue) and experiments (red) at the central wavelength of AS 3 (597.7 nm); (f) Cross-section of the interferogram from simulation (blue) and experimental data (red) at 562.57 nm.

As a demonstration of our phase retrieval algorithm, the total time delay span in our theoretical calculation was 100 fs, which suppressed the effect of phase distortion from spherical aberration induced by the displacement of translation stage. In the simulation, we solved the partial differential equations that described the coherent

153

Raman interaction between the Raman sidebands. We took all the sidebands as being far from the resonance as in Reference [21]. The equations can be written as Equation (1):

$$\frac{\partial E_q(\omega, z)}{\partial z} = -j\frac{b_0}{a_0}\beta_q \left(\int_0^\infty \chi(\Omega)\rho_{ab}(\Omega)E_{q-1}(\Omega-\omega)d\Omega + \int_0^\infty \chi^*(\Omega)\rho^*_{\ ab}(\Omega)E_{q+1}(\Omega+\omega)d\Omega \right) \quad (1)$$

where $\rho_{ab} = \frac{1}{2}\sin(\theta)\exp(j\varphi)$ and $\tan\theta = \frac{2|B|}{2\Delta\omega - D + A}, B = |B|\exp(j\varphi)$. The line profile is $\chi(\Omega) = \frac{\alpha}{\Omega - (\omega_p - \omega_s) - i\gamma}$, where α is a constant related to the spontaneous Raman cross-section, ωs and ωp are the angular frequencies of the pump and Stokes beams, and γ is the half-width of the spontaneous Raman line. The propagation constant is defined as $\beta_q(\omega) = r^*_s\hbar\omega_q N a_0$ for different wavelength and the definition of a_0 and b_0 can be found in References [8,21]. $E_q(\omega) = A_q(\omega) \times \exp(j(\varphi_q(\omega)+\omega_q t_q))$. For the amplitude $A_q(\omega)$, we used the value measured by the spectrometer. We applied the same interpolation as that in the interferogram to reconstruct $A_q(\omega)$, whose sampling points were uniformly distributed in the frequency domain. Spectral phase $\varphi_q(\omega)$ and time delay t_q of the sidebands are the fitting parameters we need to determine through simulation. The other parameters are determined by the theoretical evaluation and optimized by the calculation. In principle, there are infinite equations as n goes to $\pm\infty$. However, since our sidebands were AS3–AS5, we only considered the relevant equations, including these sidebands, within the spectral range 400 nm–700 nm. The boundary conditions are assumed to be the same as the experimental conditions. For example, in Figure 2, in addition to AS3 and AS4, the amplitudes $A_q(\omega)$ of all the other sidebands, AS2, AS5, AS6, AS7, and AS8, are zero in the initial condition.

We solved the equations using the Runge-Kutta method. In order to find the right spectral phases, we attributed the phase distortion to the dispersion of the crystal (1 mm thick synthetic single-crystal diamond in our experiment) and the retardation among the sidebands. By comparing the theoretical results (Figure 2d) with the experimental results (Figure 2c), the spectral phases were optimized by changing time delay between sidebands and the thickness of the crystal and obtained a best fit with experimental results for a thickness of 200 μm. The thickness of the crystal was about 1 mm, which was much larger than the 200 μm in the simulation. In other words, the spectral phases included the effect resulting from the dispersion of the crystal and cross phase modulation. For the results presented in the paper, we did not consider the effect of cross phase modulation separately. The reason was that when adding the cross-phase modulation induced by pump and Stokes fields into the simulation, we did not see significant improvement in the simulation. The theoretical results were also normalized in order to compare with the experimental data. Figure 2e,f displays the cross section of the interferogram at the central wavelength of AS3 at 597.7 nm AS3 and AS4 at 562.57 nm, which showed that our theoretical results fit the experimental results well. However, the results were not perfect, as we did not separately consider the phase distortions due to the nonlinear effects, such as self/cross phase modulation and spherical aberration of the mirrors.

The spectral phases retrieved in our simulation were relative. Thus, we took the spectral phase of AS3 as the reference. After finding the proper phase for AS4 using the interferogram of AS3 and AS4, we continued to retrieve the spectral phases of AS5 from the interferogram between AS4 and AS5. With the same method, the spectral phases of AS6 and AS7 were retrieved. Then we put multi-sidebands with the retrieved spectral phases together and reconstructed the corresponding interferogram to compare with the experimental result. The theoretical interferogram and experimental interferogram of multi-sidebands are shown in Figure 3. Figure 3a,b show the numerical results and experimental results with sidebands AS3, AS4, and AS5. Figure 3c,d show the numerical results and experimental results with sidebands AS3, AS4, AS5, and AS6. Figure 3e,f show the numerical results and experimental results with sidebands AS3, AS4, AS5, AS6 and AS7. After we normalized both the retrieved and the original spectrograms, we calculated the total rms deviation between simulation and experimental spectrogram as: rms $= \sqrt{\frac{1}{Nt \times N\omega} \sum \left(I_{\text{simulation}}\left(\omega, t\right) - I_{\text{experiment}}\left(\omega, t\right)\right)^2}$ to be 0.0294 for our retrieval (here ω is the frequency, t is the time delay, Nt is the total scanning steps in time domain an, $N\omega$ is the total number in frequency domain, and $I_{simulation}(\omega, t)$ and $I_{experiment}(\omega, t)$ are the normalized intensities of simulation and experiment). In Figure 4a, we show the rms deviation at a variable ω (rms $= \sqrt{\frac{1}{Nt} \sum \left(I_{\text{simulation}}\left(\omega, t\right) - I_{\text{experiment}}\left(\omega, t\right)\right)^2}$, which is a sum over scanning time steps) together with the spectral phases obtained from the retrieval. In Figure 4, we convert ω to the wavelength following $\lambda = 2\pi c/\omega$, here, c is the speed of light. The retrieval method we used did not include the iteration routine. Including the iteration in our algorithm is our future plan. The ultrafast waveform retrieved from Figure 3e,f could be found in Reference [21]. Figure 4b is the pulse retrieved from Figure 3 (b) (red), (d) (black), and (f) (blue). The pulses duration (full-width half maximum (FWHM)) is 6.06 fs for five sidebands (blue), 6.10 for four sidebands (black) and 7.37 for three sidebands (red). In principle, the five sidebands AS3, AS4, AS5, AS6, and AS7 span the spectral range from 490 nm to 597 nm and, thereby, it is possible to obtain a pulse whose full-width half-maximum (FWHM) is around 4 fs with five sidebands (5 fs with four sidebands (AS3, AS4, AS5, and AS6) and 6 fs with three sidebands (AS3, AS4, and AS5)). However, in the experiment, the spectral phases of the sidebands were distorted due to the nonlinear effect and the pulse duration (FWHM) retrieved from the numerical result (Figure 4a) is larger than that of the Fourier-transform limited pulse fs. Notably, our results showed that the pulse duration with five sidebands (AS3–7) is close to that of four sidebands (AS3–6). This is due to the fact that the spectral phase of AS7 is distorted too much and, thereby, when synthesizing the waveform using AS7, the pulse duration barely changes.

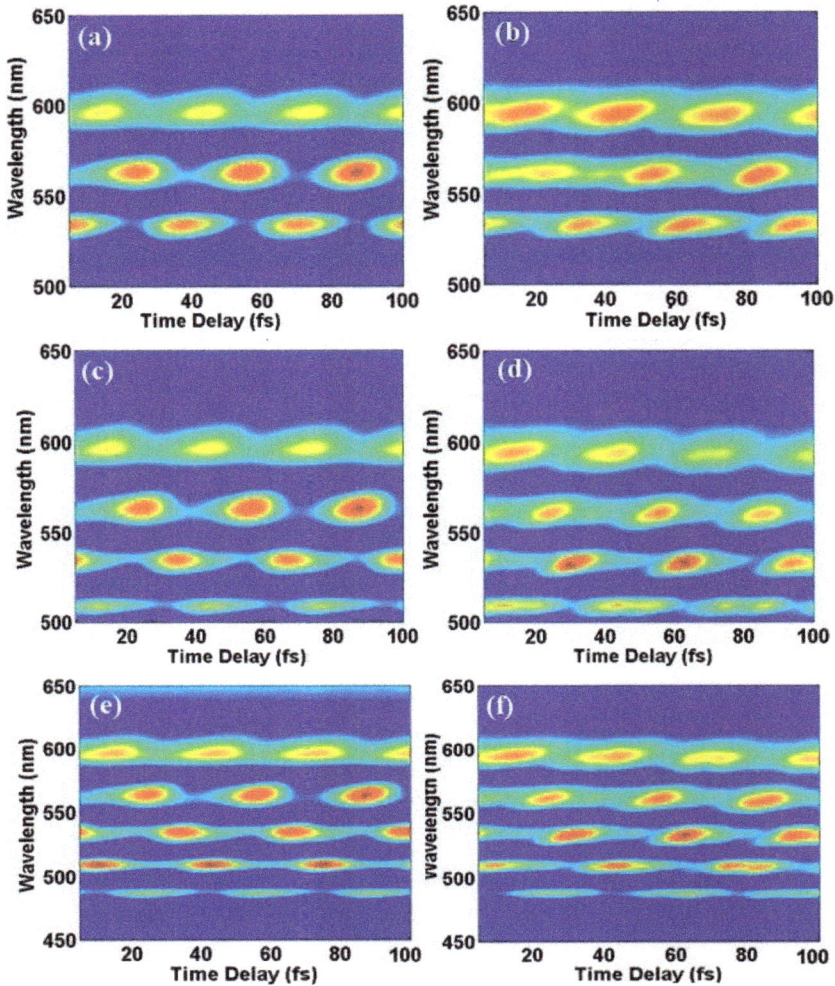

Figure 3. (a) Theoretical interferogram of AS3, AS4 and AS5; (b) Interferogram of AS3, AS4 and AS5, which is reconstructed with 2D bilinear interpolation from experimental data; (c) Theoretical interferogram of AS3, AS4, AS5, and AS6; (d) Interferogram of AS3, AS4, AS5, and AS6 reconstructed with 2D bilinear interpolation from experimental data; (e) Theoretical interferogram of AS3, AS4, AS5, AS6, and AS7; (f) Interferogram of AS3, AS4, AS5, AS6, and AS7 reconstructed with 2D bilinear interpolation from experimental result.

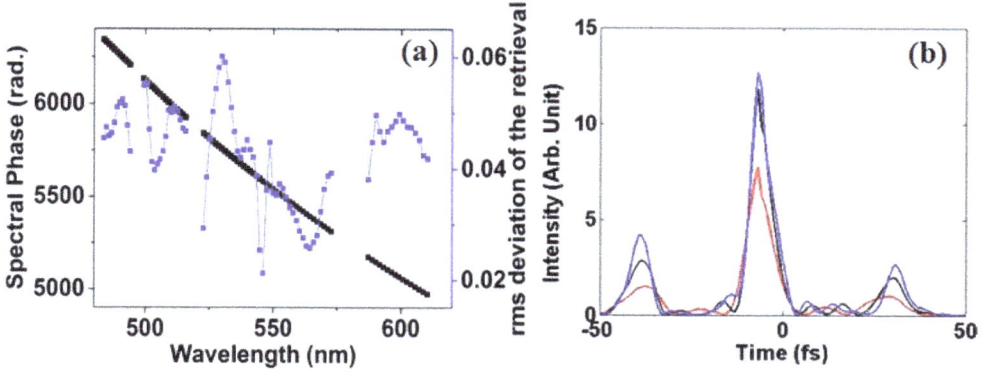

Figure 4. (a) The spectral phases retrieved from the numerical simulation (black) and the rms deviation of the simulation *vs.* experiment at different wavelength (rms is evaluated between the normalized simulated and original spectrograms by summing over the scanning of time steps). (b) The pulse retrieved from the interferogram of three sidebands AS3, AS4, and AS5 (red line), four sidebands AS3, AS4, AS5, and AS6 (black line), and five sidebands AS3, AS4, AS5, AS6, and AS7 (blue line).

4. Conclusions

In this paper, we showed the feasibility of producing an interferogram for the Raman sidebands, based on additional Raman interactions in a reflection scheme. Furthermore, using the numerical simulation, it was possible to retrieve the relative spectral phases of the Raman sidebands from the interferogram. We characterized the relative spectral phases based on the Raman nonlinear interaction. In this paper, we described the procedure of our phase retrieval algorithm, starting with the interferogram for two sidebands as an example. Then we extended it to the scenario of an interferogram obtained with more sidebands. Using the theoretical simulation, we retrieved the ultrafast pulse with five sidebands. Our interferograms were recorded in a reflection scheme. The ultrafast waveform would be produced at the focal point of the spherical mirror, even though the beams did not propagate collinearly. In principle, with a thin crystal providing sufficiently broad-band phase matching, one could still characterize the spectral phases based on the Raman interaction for collinear beams.

Acknowledgments: This work is supported by the National Science Foundation (grant No. PHY-1307153) and the Welch Foundation (grant No. A1547).

Author Contributions: Performing the experiment and numerical simulation: K. Wang, M. Zhi, and X. Hua, Drafting of manuscript: K. Wang, A. A. Zhdanova, M. Zhi, A. V. Sokolov; Critical revision: K. Wang, A. A. Zhdanova, M. Zhi, A. V. Sokolov; Planning and supervision of the research: A. V. Sokolov.

Conflicts of Interest: The authors declare no conflict of interest.

References

1. Zhao, K.; Zhang, Q.; Chini, M.; Wu, Y.; Wang, X.; Chang, Z. Tailoring a 67 attosecond pulse through advantageous phase-mismatch. *Opt. Lett.* **2012**, *37*, 3891–3893.

2. Baker, S.; Walmsley, I.A.; Marangos, J.P. Femtosecond to attosecond light pulse from a molecular modulator. *Nat. Photonics* **2011**, *5*, 664–671.

3. He, J.; Liu, J.; Kobayashi, T. Tunable Multicolored Femtosecond Pulse Generation Using Cascaded Four-Wave Mixing in Bulk Materials. *Appl. Sci.* **2014**, *4*, 444–467.

4. Wirth, A.; Hassan, M.T.; Grguraš, I.; Gagnon, J.; Moulet, A.; Luu, T.T.; Pabst, S.; Santra, R.; Alahmed, Z.A.; Azzeer, A.M.; *et al.* Synthesized Light Transients. *Science* **2011**, *334*, 195–200.

5. Motoyoshi, K.; Kida, Y.; Imasaka, T. High-Energy, Multicolor Femtosecond Pulses from the Deep Ultraviolet to the Near Infrared Generated in a Hydrogen-Filled Gas Cell and Hollow Fiber. *Appl. Sci.* **2014**, *4*, 318–330.

6. Lin, Y.Y.; Wu, P.S.; Yang, H.R.; Shy, J.T.; Kung, A.H. Broadband Continuous-Wave Multi-Harmonic Optical Comb Based on a Frequency Division-by-Three Optical Parametric Oscillator. *Appl. Sci.* **2014**, *4*, 515–524.

7. Chan, H.S.; Hsieh, Z.M.; Liang. W.H.; Kung, A.H.; Lee, C.K.; Lai, C.J.; Pan, R.P.; Peng, L.H. Synthesis and measurement of ultrafast waveforms from five discrete optical harmonics. *Science* **2011**, *331*, 1165–1168.

8. Harris, S.E.; Sokolov, A.V. Subfemtosecond pulse generation by molecular modulation. *Phys. Rev. Lett.* **1998**, *81*, 2894–2897.

9. Kien, F.L.; Liang, J.Q.; Katsuragawa, M.; Ohtsuki, K.; Hakuta, K.; Sokolov, A.V. Subfemtosecond pulse generation with molecular coherence control in stimulated Raman scattering. *Phys. Rev. A* **1999**, *50*, 1562–1571.

10. Yoshikawa, S.; Imasaka, T. A new approach for the generation of ultrashort optical pulses. *Opt. Commun.* **1993**, *96*, 94–98.

11. Suzuki, T.; Hirai, M.; Katsuragawa, M. Octave-Spanning Raman Comb with Carrier Envelope Offset Control. *Phys. Rev. Lett.* **2008**, *101*.

12. Couny, F.; Benabid, F.; Roberts, P.J.; Light, P.S.; Raymer, M.G. Generation and photonic guidance of multi-octave optical-frequency combs. *Science* **2007**, *318*, 1118–1121.

13. Yan, H.; Strickland, D. Effect of Two-Photon Stark Shift on the Multi-Frequency Raman Spectra. *Appl. Sci.* **2014**, *4*, 390–401.

14. Gold, D.C.; Weber, J.J.; Yavuz, D.D. Continuous-Wave Molecular Modulation Using a High-Finesse Cavity. *Appl. Sci.* **2014**, *4*, 498–514.

15. Abdolvand, A.; Walser, A.M.; Ziemienczuk, M.; Nguyen, T.; Russell, P.S.J. Generation of a phase-locked Raman frequency comb in gas-filled hollow-core photonic crystal fiber. *Opt. Lett.* **2012**, *37*, 4362–4364.

16. Sokolov, A.V.; Harris, S.E. Ultrashort pulse generation by molecular modulation. *J. Opt. B Quantum Semiclass. Opt.* **2003**, *5*.

17. Zhi, M.; Wang, K.; Hua, X.; Sokolov, A.V. Pulse-Shaper-Assisted phase control of a coherent broadband spectrum of Raman sidebands. *Opt. Lett.* **2011**, *36*, 4032–4034.

18. Zhi, M.; Wang, K.; Hua, X.; Strycker, B.D.; Sokolov, A.V. Shaper-Assisted phase optimization of a broad "holey" spectrum. *Opt. Express* **2011**, *19*, 23400–23407.

19. Strohaber, J.; Zhi, M.; Sokolov, A.V.; Kolomenskii, A.A.; Paulus, G.G.; Schuessler, H.A. Coherent transfer of optical orbital angular momentum in multi-order Raman sideband generation. *Opt. Lett.* **2012**, *37*, 3411–3413.

20. Zhi, M.; Wang, K.; Hua, X.; Schuessler, H.; Strohaber, J.; Sokolov, A.V. Generation of femtosecond optical vortices by molecular modulation in a Raman-active crystal. *Opt. Express* **2013**, *21*, 27750–27758.

21. Wang, K.; Zhi, M.; Hua, X.; Sokolov, A.V. Ultrafast waveform synthesis and characterization using coherent Raman sidebands in a reflection scheme. *Opt. Express* **2014**, *22*, 21411–21420.

22. Zhi, M.; Wang, X.; Sokolov, A.V. Broadband coherent light generation in diamond driven by femtosecond pulses. *Opt. Express* **2008**, *16*, 12139–12147.

23. Zhi, M.; Sokolov, A.V. Broadband coherent light generation in a Raman active crystal driven by two-color femtosecond laser pulses. *Opt. Lett.* **2007**, *32*, 2251–2253.

24. Boyd, R.W.; Fisher, G.L. Nonlinear Optical Materials. In *Encyclopedia of Materials: Science and Technology*; Elsevier Science Ltd.: Oxford, UK, 2001.

A Simple Method for the Evaluation of the Pulse Width of an Ultraviolet Femtosecond Laser Used in Two-Photon Ionization Mass Spectrometry

Tomoko Imasaka, Akifumi Hamachi, Tomoya Okuno and Totaro Imasaka

Abstract: A simple method was proposed for on-site evaluation of the pulse width of an ultraviolet femtosecond laser coupled with a mass spectrometer. This technique was based on measurement of a two-photon ionization signal in mass spectrometry by translation of the prism in the pulse compressor of the femtosecond laser. The method was applied to optical pulses that were emitted at wavelengths of 267, 241, and 219 nm; the latter two pulses were generated by four-wave Raman mixing using the third harmonic emission of a Ti:sapphire laser (267 nm) in hydrogen gas. The measurement results show that this approach is useful for evaluation of the pulse width of the ultraviolet femtosecond laser used in mass spectrometry for trace analysis of organic compounds.

Reprinted from *Appl. Sci.*. Cite as: Imasaka, T.; Hamachi, A.; Okuno, T.; Imasaka, T. A Simple Method for the Evaluation of the Pulse Width of an Ultraviolet Femtosecond Laser Used in Two-Photon Ionization Mass Spectrometry. *Appl. Sci.* **2016**, *6*, 136.

1. Introduction

Ultrashort laser pulses have been used successfully in a variety of applications including trace analysis of organic compounds [1]. Several techniques have been developed to measure the pulse width of the femtosecond laser, including autocorrelation (AC), spectral phase interferometry for direct electric field reconstruction (SPIDER), and frequency-resolved optical gating (FROG) [2]. Among these methods, the FROG technique is widely used and has several variations in its implementation. For example, second harmonic generation (SHG), polarization gating (PG), or self-diffraction (SD) can be used for the nonlinear optical effect. To expand the spectral domain to measure shorter pulse widths, third harmonic generation (THG), cross-correlation (X), and four-wave mixing (FWM) have all been used to date [3–6]. On the other hand, the pulse width of the UV femtosecond laser has been measured based on autocorrelation using a sold material, e.g., a CaF_2 plate, as a nonlinear optical device [7–10]. Moreover, the vacuum-ultraviolet (VUV) laser pulse has been measured by focusing it with a near-infrared (NIR) probe pulse into xenon to observe the cross-correlation signal of ionization [11–15]. However, the pulse duration would be changed by traveling even in ambient air from the location of the device used to perform the pulse width measurement to the point of application. Complex techniques such as attosecond streaking (AS) or FROG for complete reconstruction of attosecond bursts (FROG-CRAB) can be used to measure the pulse width in a vacuum chamber used for extreme ultraviolet (EUV) pulse generation [16,17]. However, these techniques are rather complicated for use in practical applications.

For simple on-site measurement of pulse widths, a device based on fringe-resolved autocorrelation (FRAC) was developed, which consisted of an interferometer and a time-of-flight mass spectrometer that was used as a two-photon-response detector to measure a non-resonant two-photon ionization signal [18–21]. However, the laser beam must be split into two parts and recombined after the interferometer, which reduces the laser pulse energy and then the sensitivity of the mass spectrometer. Also, the low-dispersion aluminum mirrors used in the interferometric system have substantial reflection losses in the deep-ultraviolet (DUV) region. In addition, the mass spectrometer must be operated for long periods, e.g., for periods of days, without maintenance to compare the data from repeated measurements obtained on the same day. Thus, it requires sufficient optical system stability for the pulse width measurements to be performed. In fact, we tried to measure an autocorrelation trace for a DUV pulse generated through several nonlinear optical processes combined in series, but the attempt was unsuccessful due to low pulse energy, poor beam quality, and instability of the laser pulse energy although performance of the laser was sufficient for application to mass spectrometry. Therefore, a simple and rugged instrument that does not contain an interferometer is highly desirable for evaluation of the widths of pulses from DUV femtosecond lasers. On the other hand, DUV femtosecond lasers have been used successfully for two-photon ionization applications in mass spectrometry [1]. Because the dispersion caused by the optical components or even by the ambient air cannot be negligible in the DUV region, a device such as a prism pair must be used for pulse compression in the laser system to generate a nearly-transform-limited pulse in the mass spectrometer [22]. It should be noted that a technique referred to as multiphoton intrapulse interference phase scan (MIIPS) or dispersion scan (d-can) has been reported, in which the SHG spectrum is measured by changing the dispersion in the beam path, e.g., by translating a wedge or prism, or changing the angle of a grating [23–29]. This method is useful for observing a two-dimensional display of the delay (dispersion) and the spectrum, allowing the full characterization of the chirp of the pulse. However, the use of a crystal for SHG limits the spectral range to the NIR-visible region and also makes the on-site measurement of the pulse width difficult.

In this study, we propose a simple method to evaluate the widths of pulses from a DUV femtosecond laser without use of an interferometer. This technique is based on measurement of a non-resonant two-photon ionization signal in mass spectrometry by translating the prism in the laser's pulse compressor. This method was applied to femtosecond optical pulses emitted at 241 and 219 nm, in addition to pulses emitted at 267 nm, which were generated by four-wave Raman mixing in hydrogen gas. The results obtained herein were compared with the transform-limited pulse widths that were calculated from the spectral bandwidths of the laser used in the experiments.

2. Theoretical Calculations

The electric field and intensity characteristics of an optical pulse with respect to time (t) can be assumed to have Gaussian profiles [2].

$$\tilde{E}(t) = E_0 e^{-at^2} e^{i(\omega t + bt^2)} \tag{1}$$

$$I(t) = |E_0|^2 e^{-2at^2} \tag{2}$$

where "a" is related to the pulse duration, Δt, as shown below, and "b" is a chirp parameter.

$$\sqrt{\frac{2\ln 2}{a}} \approx \Delta t \tag{3}$$

The second-order FRAC signal can be expressed as

$$I^2_{FRAC}(\tau) = \int_{-\infty}^{\infty} \left| \left(\tilde{E}(t - \tau) + \tilde{E}(t) \right)^2 \right|^2 dt \tag{4}$$

where τ is the time delay between the two pulses in the interferometer [21]. This parameter can be set to zero because the beam is not separated in this study (i.e., no interferometer is used), and this leads to Equation (5).

$$I^2_{FRAC}(\tau = 0) = \int_{-\infty}^{\infty} \left| \left(2\tilde{E}(t) \right)^2 \right|^2 dt = 16 \int_{-\infty}^{\infty} \left| \left(\tilde{E}(t) \right)^2 \right|^2 dt = 16|E_0|^4 \int_{-\infty}^{\infty} e^{-4at^2} dt = 16A^2 \sqrt{\frac{a}{\pi}} \tag{5}$$

where the parameter, A, is defined as a pulse energy by the following equation.

$$\int_{-\infty}^{\infty} I(t)dt = |E_0|^2 \int_{-\infty}^{\infty} e^{-2at^2} dt = A \tag{6}$$

When the signal intensity becomes one half of this value via a chirp of the pulse,

$$I^2_{F-L}(\tau = 0) = 2I^2_{CP}(\tau = 0) \tag{7}$$

162

where I^2_{FTL} and I^2_{CP} are the signal intensities of the transform-limited and chirped pulses, respectively. The following equation is then obtained.

$$\sqrt{a_{FTL}} = 2\sqrt{a_{CP}} \tag{8}$$

where a_{FTL} and a_{CP} are the parameters that were calculated using Equation (3). Then,

$$\Delta t_{CP} = 2\Delta t_{FTL} \tag{9}$$

where Δt_{FTL} and Δt_{CP} are the pulse durations of the transform-limited and chirped pulses, respectively. The relationship between Δt_{CP} and Δt_{FTL} can be expressed as Equation (10) [21].

$$\Delta t_{CP} = \sqrt{(\Delta t_{FTL})^2 + \left(\frac{\varphi_2 \times 4\ln 2}{\Delta t_{FTL}}\right)^2} \tag{10}$$

where φ_2 is the group delay dispersion (GDD), which can be expressed as a product of the group velocity dispersion (GVD) determined using the Sellmeier equation [30,31] and the effective length of the optical material in the beam path, $\ell - \ell_0$ (see Figure 1a). Then,

$$\varphi_2 = GDD = GVD \times 2 \times (\ell - \ell_0) \tag{11}$$

where $\ell = (2 \tan 34°) \times L$ for a Brewster prism used at approximately 250 nm. The factor of 2 in Equation (11) is multiplied due to a double pass of the beam in the prism. When the dispersion is canceled by a prism pair, $GDD = 0$ and $\ell_0 = (2 \tan 34°) \times L_0$ where ℓ_0 and L_0 are the parameters of ℓ and L at which the dispersion is canceled (the two-photon ionization signal is maximal). A parameter of X can be defined as a displacement of the prism position from the optimum location that cancels the dispersion (i.e., $GDD = 0$), then $X \equiv L - L_0$. When the intensity of the two-photon ionization signal decreases to one half of the maximum value, $L = L_{1/2}$ and then $X_{1/2} = L_{1/2} - L_0$. From Equations (9) and (10), Δt_{FTL} can then be rewritten as

$$\Delta t_{FTL} = \frac{\sqrt{4\ln 2}}{\sqrt[4]{3}} \sqrt{\varphi_2\left(X_{1/2}\right)} \tag{12}$$

where $\varphi_2(X_{1/2})$ is the value of GDD at $X = X_{1/2}$. This equation suggests that the transform-limited pulse width can be calculated by measuring the parameter of $X_{1/2}$ in the experiment.

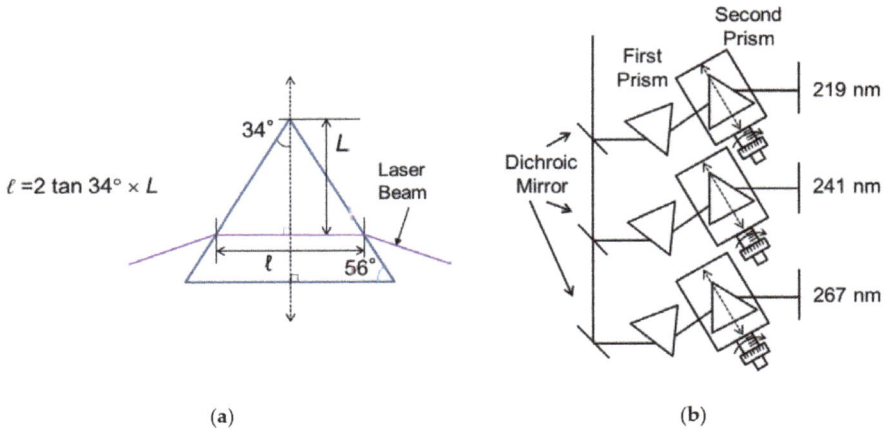

Figure 1. (a) Prism parameters; (b) configuration of the prisms in a pulse compressor. Dotted arrows show the direction of displacement of the prisms. The parameters, ℓ and L, are the path length of the laser beam in the prism and the distance from the top of the prism to the laser beam, respectively. The angle of the prism is specified in the figure. Three prisms in (b) can be manually translated by rotating the nobs of the differential micrometers independently.

The actual pulse width is, however, broadened by several reasons: (1) an initial laser pulse would be neither Gaussian-shaped nor transform-limited; (2) the laser pulse would be chirped even at the optimal prism position because of the third-order dispersion (TOD) of the fused silica of the prisms used for pulse compression; and (3) the pulse front would be deformed by mode-change and self-focusing during beam transmission in a hollow capillary filled with hydrogen gas (many hot spots were observed in the beam pattern). In order to take account of the deterioration of the laser beam, a parameter, α, can be introduced into Equation (12).

$$\Delta t_{REAL} = \frac{\sqrt{4\ln 2}}{\sqrt[4]{3}} \sqrt{\alpha \times \varphi_2(X_{1/2})} \tag{13}$$

where Δt_{REAL} is the pulse width observed in the experiment and α is the parameter showing the deviation from the transform-limited pulse. When TOD is only a factor responsible for the distortion of the pulse, α can be written as $1 + TOD/GVD$ where

$$GVD = \frac{\lambda^3}{2\pi c^2} \frac{d^2 n}{d\lambda^2} \text{ and } TOD = -\left(\frac{\lambda}{2\pi c}\right)^2 \frac{1}{c} \left(3\lambda^2 \frac{d^2 n}{d\lambda^2} + \lambda^3 \frac{d^3 n}{d\lambda^3}\right) \tag{14}$$

Equations (11)–(13) lead to the following equations.

$$\text{When } \alpha = 1, \ \Delta t_{\text{FTL}} = \frac{\sqrt{16\ln 2 \tan 34°}}{\sqrt[4]{3}} \sqrt{GVD \times X_{1/2}} \tag{15}$$

$$\text{When } \alpha > 1, \ \Delta t_{\text{REAL}} = \frac{\sqrt{16\ln 2 \tan 34°}}{\sqrt[4]{3}} \sqrt{GVD \times \alpha \times X_{1/2}} = \Delta t_{\text{FTL}} \sqrt{\alpha} \tag{16}$$

As shown in Equation (16), the value of $X_{1/2}$ to be obtained for a transform-limited pulse is actually expanded to a value of $\alpha X_{1/2}$ in the experiment by some undesirable effects arising from additional chirps and deterioration of the laser pulse. The graph showing the relationship between Δt_{REAL} and $\alpha X_{1/2}$ is the same as that of Δt_{FTL} vs. $X_{1/2}$, as shown in Equations (15) and (16). As can be recognized from Equations (15) and (16), Δt_{REAL} is expanded by a factor of $\sqrt{\alpha}$ from Δt_{FTL} that can be calculated from the spectral bandwidth of the laser beam. The parameter, α, can be calculated from the ratio of Δt_{REAL} and Δt_{FTL}, suggesting a degree of the deviation from the ideal transform-limited Gaussian pulse.

A train of ultrashort optical pulses is generated by superposition of the laser emissions, which are phase-locked to each other. In this study, three beams, i.e., 9ω (267 nm), 10ω (241 nm), and 11ω (219 nm), where $\omega = 4155$ cm^{-1}, were superimposed and were then used as one of the test beams. The spectral width, $\Delta\omega_n$, and the parameter, a_n, where $n = 1, 2, 3$, for each beam, were assumed to be $\Delta\omega_1 = \Delta\omega_2 = \Delta\omega_3 = \Delta\omega$ and $a_1 = a_2 = a_3 = a$ for the purposes of this study. A train of ultrashort pulses is formed under these conditions if the three emission lines are phase locked. Thus, the phase locking expected to occur during the process of four-wave Raman mixing, can be confirmed by comparing the experimental data with the simulation results. The GDD can then be expressed as follows [2,21].

$$\varphi_2 \equiv GDD = \frac{b}{2(a^2 + b^2)} \tag{17}$$

$$\frac{a}{2(a^2 + b^2)} = \ln 2\left(\frac{2}{\Delta\omega}\right)^2, b = \frac{(\Delta\omega)^2 \times GDD}{2(\Delta t)^2} \tag{18}$$

When the laser emissions are phase-locked together, $I^2_{\text{FRAC}} (\tau = 0)$ can be written as

$$I^2_{\text{FRAC}}(\tau = 0) = \int_{-\infty}^{\infty} \left| \left(2(\tilde{E}_1(t) + \tilde{E}_2(t) + \tilde{E}_3(t))\right)^2 \right|^2 dt \tag{19}$$

By assuming that $E_{0,1} = E_{0,2} = E_{0,3} = 1$ for simplicity,

$$I^2_{\text{FRAC}}(\tau = 0) = 16 \int_{-\infty}^{\infty} \left| \left(e^{-a_1 t^2} e^{i(\omega_1 t + b_1 t^2)} + e^{-a_2 t^2} e^{i(\omega_2 t + b_2 t^2)} + e^{-a_3 t^2} e^{i(\omega_3 t + b_3 t^2)}\right)^2 \right|^2 dt \tag{20}$$

It should be noted that the three transform-limited pulses are assumed to be superimposed in-phase without any chirp in Equation (20). In contrast, when the laser emissions are not phase-locked to each other (*i.e.*, they are randomly phased), I^2_{FRAC} ($\tau = 0$) can then be expressed as

$$I^2_{FRAC}(\tau = 0) = \int_{-\infty}^{\infty} \left| \left(2e^{-a_1t^2}e^{i(\omega_1t+b_1t^2)}\right)^2 \right|^2 dt + \int_{-\infty}^{\infty} \left| \left(2e^{-a_2t^2}e^{i(\omega_2t+b_2t^2)}\right)^2 \right|^2 dt + \int_{-\infty}^{\infty} \left| \left(2e^{-a_3t^2}e^{i(\omega_3t+b_3t^2)}\right)^2 \right|^2 dt \quad (21)$$

The spectral width can be assumed to be $\Delta\omega = 2\omega$, although the actual spectral shape is far beyond a Gaussian profile. The experimental data can then be compared with the data calculated using Equations (20) and (21).

3. Experimental

An optical parametric amplifier (OPA, OPerA-Solo, <50 fs, Coherent, Inc., Santa Clara, CA, USA) was pumped by a Ti:sapphire laser (800 nm, 35 fs, 4 mJ, 1 kHz, Elite, Coherent, Santa Clara, CA, USA). The beam of the Ti:sapphire laser that remained for mixing with the OPA output was used for third harmonic generation (267 nm). The remaining fundamental beam (800 nm) and the signal beam of the OPA (1200 nm) were spatially and temporally superimposed on each other and were focused into a hollow capillary filled with hydrogen gas for vibrational molecular modulation. The third harmonic emission (267 nm) was then focused into the hollow capillary to provide frequency modulation and generate vibrational Raman emissions at 241 and 219 nm. The three-color beam (267, 241, 219 nm) was separated into three individual beams using three dielectric mirrors (Sigma Koki, Tokyo, Japan) placed in series, as shown in Figure 1b. The laser beams passed through three pairs of prisms for pulse compression. The beams were then reflected by a pair of roof mirrors and were recombined into a single beam using the dielectric mirrors. The three-color beam was then focused by a concave mirror into a molecular beam in a mass spectrometer (Hikari Gijyutsu Corp., Fukuoka, Japan). The energy of each pulse measured in front of the mass spectrometer was *ca.* 1 µJ. In the experiment, 1,4-dioxane was introduced into the mass spectrometer for recording of a non-resonant two-photon ionization signal [20]. The signal intensity was measured by translating the second prism to alter the positive dispersion in the compressor. The parameter $X_{1/2}$ was determined to be the half width at half maximum of the observed data. The spectral bandwidth of the laser was measured using a spectrometer (Maya2000pro, spectral resolution 1.5 nm, Ocean Optics, Dunedin, FL, USA), the resolution of which was calibrated using a mercury lamp (Ocean Optics, Dunedin, FL, USA) at 254 nm.

4. Results and Discussion

4.1. Calculations

Figure 2 shows the dependence of Δt_{REAL} on $\alpha X_{1/2}$ (or Δt_{FTL} on $X_{1/2}$), which was calculated in the range from 2ω to 16ω ($\omega = 4155$ cm^{-1}) using Equation (16). Because the dispersion increases at higher frequencies, the pulse width increases even at the same value of $\alpha X_{1/2}$. For example, when $\alpha X_{1/2} = 1$ mm, the pulse width becomes 5.0 fs at 2ω and 95 fs at 16ω. Another example would be $\alpha X_{1/2} = 10$ mm, providing pulse widths of 16 fs at 2ω and 300 fs at 16ω. Therefore, the pulse width can be evaluated based on the observed data to show the dependence of the two-photon ionization signal on the displacement (X) of the second prism in the pulse compressor.

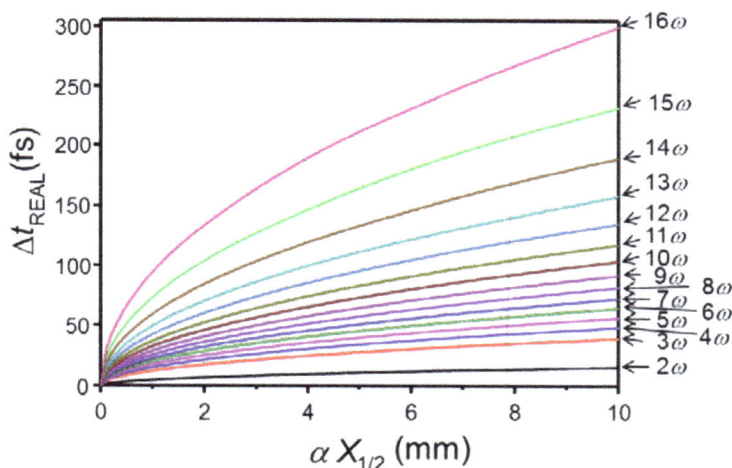

Figure 2. Calculated dependence of the parameter, Δt_{REAL}, on the parameter, $\alpha X_{1/2}$. The laser frequency, $n\omega$, is shown in the figure, where n and ω are the order of Raman sidebands and the Raman shift frequency of molecular hydrogen ($\omega = 4155$ cm^{-1}), respectively. The definitions of the parameters, Δt_{REAL} and $\alpha X_{1/2}$, are given in the text.

4.2. Puse Width Evaluation

Figure 3 shows the dependence of the signal intensities measured as a function of the prism displacement, X, for laser beams emitted at 9ω (=267 nm), 10ω (=241 nm), and 11ω (=219 nm). Because $\alpha X_{1/2} = 4.9$ mm at 9ω, the pulse width, Δt_{REAL}, can be calculated to be 64 fs from the data shown in Figure 2. For the values of $\alpha X_{1/2} = 5.6$ mm at 10ω and 1.7 mm at 11ω, the corresponding values of Δt_{REAL} can be calculated to be 78 and 49 fs, respectively. The spectral bandwidth was

167

measured using a spectrometer, and the estimated values were *ca.* 3.3, 2.9, and 2.5 nm after calibration of the resolution (1.5 nm) of the spectrometer at 9ω, 10ω, and 11ω, respectively. From the relationship of $\Delta t \times \Delta v \geqslant k$ (where $k = 0.441$ for a Gaussian pulse), $\underline{\Delta t}_{FTL}$ can be calculated to be 36, 34, and 35 fs at 9ω, 10ω, and 11ω, respectively. The parameter, α, can then be calculated to *ca.* 3.2, 5.3, and 2.0 for 9ω, 10ω, and 11ω, respectively. These results suggest that the anti-Stokes beam has the shortest pulse width among them and has a slightly (1.4 times) larger pulse width than the width expected for a transform-limited pulse.

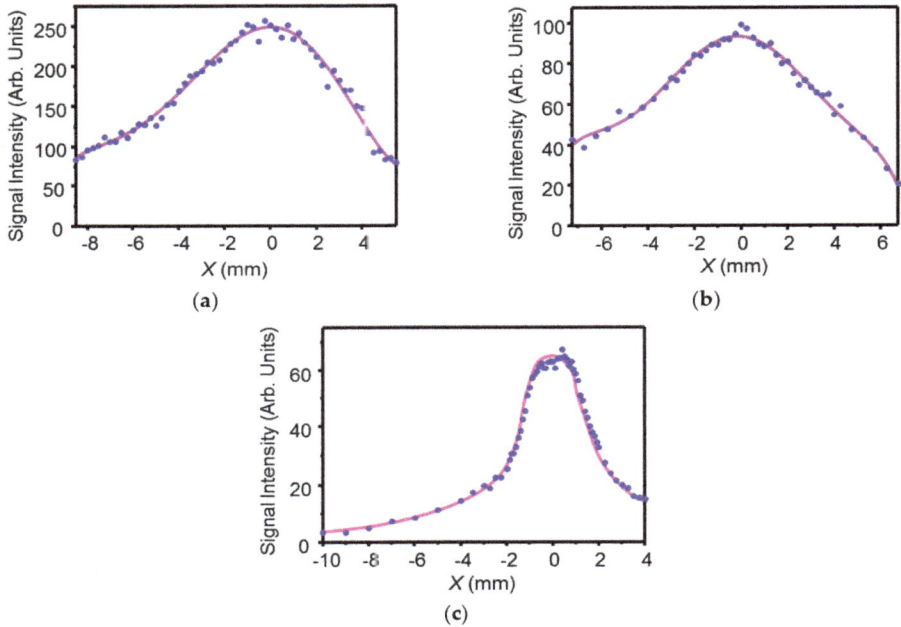

Figure 3. Observed dependence of the signal intensity, $I^2{}_{FRAC}$ ($\tau = 0$), on the parameter, X. The dots in the figure are the observed data, and a solid curve is a guide to the eye. The parameter of $X_{1/2}$ can be obtained by calculating the half width at the half maximum of the signal peak. Laser operating wavelengths: (a) 267; (b) 241; and (c) 219 nm.

While the spectral profile of the three-color beam composed from the three spectral lines (267, 241, 219 nm) was far beyond the Gaussian shape, the presented technique was applied to this type of beam on a trial basis. The predicted results were calculated for two cases: (1) where the phases of the emissions are random; and (2) where the emission lines are phase-locked. In the former case, a broad band profile was obtained, as shown in Figure 4a, in which $\alpha X_{1/2} = 2.8$ mm. In the latter case, a very sharp peak, *i.e.*, $\alpha X_{1/2} = 0.015$ mm, was obtained, as shown in Figure 4b.

The experimental data shown in Figure 4c consisted of two components, where one is a sharp peak ($\alpha X_{1/2} = 0.12$ mm) at the center, and the other is a broad band observed as a pedestal. It should be noted that the signal intensity was highly sensitive to the positions of the second prisms in the vicinity of the maximum value, and that this result was obtained by carefully translating these prisms simultaneously after critical optimization of their positions.

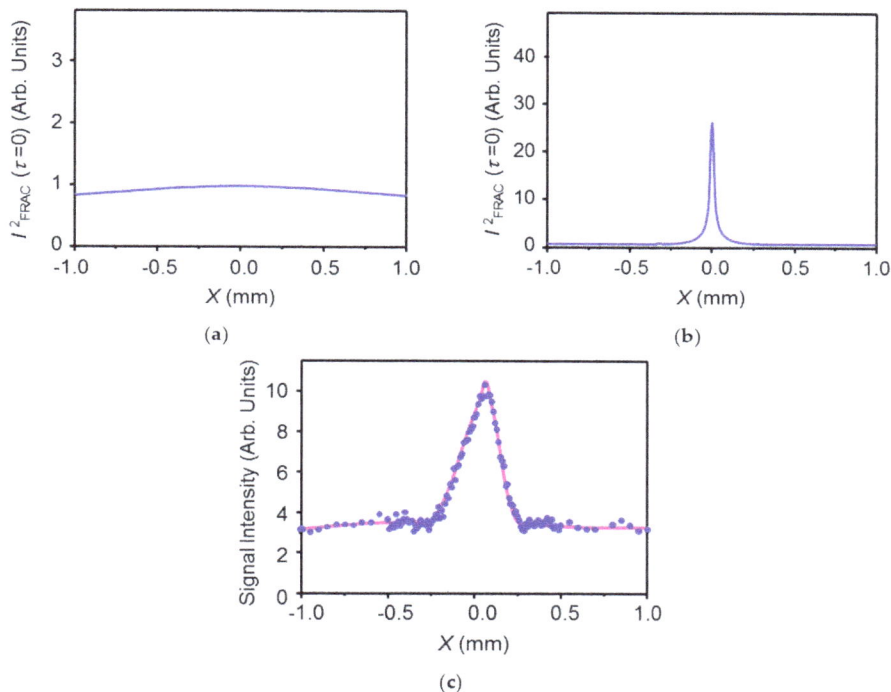

Figure 4. Calculated dependence of the parameter, I^2_{FRAC} ($\tau = 0$), on the parameter, X: (**a**) random phase; and (**b**) phase-locked; (**c**) Observed dependence of the signal intensity, I^2_{FRAC} ($\tau = 0$), on the parameter, X. The parameter of $X_{1/2}$ can be obtained by calculating the half width at the half maximum of the signal peak. In this case, a three-color beam emitting at 267, 241, and 219 nm was introduced into the molecular beam in the mass spectrometer.

It is possible to consider that a sharp peak would arise from the phase-locked components while another would arise from the phase-random components. The former peak width (0.12 mm), corresponding to a pulse width of 11 fs, was apparently broader than the theoretically predicted value (0.015 mm), corresponding to a pulse width of 4.0 fs. This discrepancy is likely to arise from insufficient precision in phase control of the three emissions, during which the prisms were moved simultaneously using translation stages equipped with manually driven differential micrometers (see Figure 1b). Another explanation for the discrepancy, and the appearance of the pedestal, could be the spectral phase and amplitude fluctuations between the three pulses used in this study [32].

5. Conclusions

A simple method for evaluation of the widths of pulses from a UV femtosecond laser was proposed that was based on measurement of a two-photon ionization signal by translation of the second prism in the laser's pulse compressor. The pulse widths that were observed experimentally showed reasonably good agreement with the values that were calculated from the spectral bandwidths, which were measured using a spectrometer. The method presented here could be applied to lasers with shorter pulse widths by using a prism with lower TOD, e.g., a prism made of CaF_2 or MgF_2. The main advantage of this method is the minimal loss of pulse energy during measurement of the pulse width. Therefore, this technique can be used to evaluate the widths of pulses from the UV femtosecond laser that is used as an ionization source in mass spectrometry for practical trace analysis.

Acknowledgments: This research was supported by Grants-in-Aid from the Japan Society for the Promotion of Science (JSPS) KAKENHI (Grant Nos. 23245017, 24510227, 26220806, and 15K01227).

Author Contributions: Totaro Imasaka conceived and designed the experiments and theory; Tomoko Imasaka designed the theory; Akifumi Hamachi and Tomoya Okuno performed the experiments; All authors analyzed the data; Totaro Imasaka contributed reagents/materials/analysis tools; Totaro Imasaka and Tomoko Imasaka wrote the paper.

Conflicts of Interest: The authors declare no conflict of interest.

References

1. Imasaka, T. Gas chromatography/multiphoton ionization/time-of-flight mass spectrometry using a femtosecond laser. *Anal. Bioanal. Chem.* **2013**, *405*, 6907–6912.
2. Trebino, R. *Frequency-Resolved Optical Gating: The Measurement of Ultrashort Laser Pulses;* Kluwer Academic Publishers: Boston, MA, USA, 2002.
3. Chadwick, R.; Spahr, E.; Squier, J.A.; Durfee, C.G.; Walker, B.C.; Fittinghoff, D.R. Fringe-free, background-free, collinear third-harmonic generation frequency-resolved optical gating measurements for multiphoton microscopy. *Opt. Lett.* **2006**, *31*, 3366–3368.
4. Meshulach, D.; Barad, Y.; Silberberg, Y. Measurement of ultrashort optical pulses by third-harmonic generation. *J. Opt. Soc. Am. B* **1997**, *14*, 2122–2125.
5. Nomura, Y.; Shirai, H.; Ishii, K.; Tsurumachi, N.; Voronin, A.A.; Zheltikov, A.M.; Fuji, T. Phase-stable sub-cycle mid-infrared conical emission from filamentation in gases. *Opt. Express* **2012**, *20*, 24741–24747

6. Shverdin, M.Y.; Walker, D.R.; Yavuz, D.D.; Yin, G.Y.; Harris, S.E. Generation of a single-cycle optical pulse. *Phys. Rev. Lett.* **2005**, *94*, 033904.

7. Tzankov, P.; Steinkellner, O.; Zheng, J.; Mero, M.; Freyer, W.; Husakou, A.; Babushkin, I.; Herrmann, J.; Noack, F. High-power fifth-harmonic generation of femtosecond pulses in the vacuum ultraviolet using a Ti:sapphire laser. *Opt. Express* **2007**, *15*, 6389–6395.

8. Homann, C.; Lang, P.; Riedle, E. Generation of 30 fs pulses tunable from 189 to 240 nm with an all-solid-state setup. *J. Opt. Soc. Am. B* **2012**, *29*, 2765–2769.

9. Zuo, P.; Fuji, T.; Horio, T.; Adachi, S.; Suzuki, T. Simultaneous generation of ultrashort pulses at 158 and 198 nm in a single filamentation cell by cascaded four-wave mixing in Ar. *Appl. Phys. B* **2012**, *108*, 815–819.

10. Mero, M.; Zheng, J. Femtosecond optical parametric converter in the 168–182-nm range. *Appl. Phys. B* **2012**, *106*, 37–43.

11. Trushin, S.A.; Fuss, W.; Kosma, K.; Schmid, W.E. Widely tunable ultraviolet sub-30-fs pulses from supercontinuum for transient spectroscopy. *Appl. Phys. B* **2006**, *85*, 1–5.

12. Trushin, S.A.; Kosma, K.; Fuß, W.; Schmid, W.E. Sub- 10-fs supercontinuum radiation generated by filamentation of few-cycle 800 nm pulses in argon. *Opt. Lett.* **2007**, *32*, 2432–2434.

13. Kosma, K.; Trushin, S.A.; Schmid, W.E.; Fuß, W. Vacuum ultraviolet pulses of 11 fs from fifth-harmonic generation of a Ti:sapphire laser. *Opt. Lett.* **2008**, *33*, 723–725.

14. Beutler, M.; Ghotbi, M.; Noack, F. Generation of intense sub-20-fs vacuum ultraviolet pulses compressed by material dispersion. *Opt. Lett.* **2011**, *36*, 3726–3728.

15. Ghotbi, M.; Trabs, P.; Beutler, M.; Noack, F. Generation of tunable sub-45 femtosecond pulses by noncollinear four-wave mixing. *Opt. Lett.* **2013**, *38*, 486–488.

16. Wirth, A.; Hassan, M.T.; Grguraš, I.; Gagnon, J.; Moulet, A.; Luu, T.T.; Pabst, S.; Santra, R.; Alahmed, Z.A.; Azzeer, A.M.; *et al.* Synthesized light transients. *Science* **2011**, *334*, 195–200.

17. Mairesse, Y.; Quéré, F. Frequency-resolved optical gating for complete reconstruction of attosecond bursts. *Phys. Rev. A* **2005**, *71*, 011401.

18. Zaitsu, S.; Miyoshi, Y.; Kira, F.; Yamaguchi, S.; Uchimura, T.; Imasaka, T. Interferometric characterization of ultrashort deep ultraviolet pulses using a multiphoton ionization mass spectrometer. *Opt. Lett.* **2007**, *32*, 1716–1718.

19. Imasaka, T.; Imasaka, T. Searching for a molecule with a wide frequency domain for non-resonant two-photon ionization to measure the ultrashort optical pulse width. *Opt. Commun.* **2012**, *285*, 3514–3518.

20. Okuno, T.; Imasaka, T.; Kida, Y.; Imasaka, T. Autocorrelator for measuring an ultrashort optical pulse width in the ultraviolet region based on two-photon ionization of an organic compound. *Opt. Commun.* **2014**, *310*, 48–52.

21. Imasaka, T.; Okuno, T.; Imasaka, T. The search for a molecule to measure an autocorrelation trace of the second/third harmonic emission of a Ti:sapphire laser based on two-photon resonant excitation and subsequent one-photon ionization. *Appl. Phys. B* **2013**, *113*, 543–549.

22. Hamachi, A.; Okuno, T.; Imasaka, T.; Kida, Y.; Imasaka, T. Resonant and nonresonant multiphoton ionization processes in the mass spectrometry of explosives. *Anal. Chem.* **2015**, *87*, 3027–3031.

23. Lozovoy, V.V.; Pastirk, I.; Dantus, M. Multiphoton intrapulse interference. IV. Ultrashort laser pulse spectral phase characterization and compensation. *Opt. Lett.* **2004**, *29*, 775–777.

24. Xu, B.; Gunn, J.M.; Dela Cruz, J.M.; Lozovoy, V.V.; Dantus, M. Quantitative investigation of the multiphoton intrapulse interference phase scan method for simultaneous phase measurement and compensation of femtosecond laser pulses. *J. Opt. Soc. Am. B* **2006**, *23*, 750–759.

25. Coello, Y.; Lozovoy, V.V.; Gunaratne, T.C.; Xu, B.; Borukhovich, I.; Tseng, C.-H.; Weinacht, T.; Dantus, M. Interference without an interferometer: A different approach to measuring, compressing, and shaping ultrashort laser pulses. *J. Opt. Soc. Am. B* **2008**, *25*, A140–A150.

26. Lozovoy, V.V.; Xu, B.; Coello, Y.; Dantus, M. Direct measurement of spectral phase for ultrashort laser pulses. *Opt. Express* **2008**, *16*, 592–597.

27. Miranda, M.; Fordell, T.; Arnold, C.; L'Huillier, A.; Crespo, H. Simultaneous compression and characterization of ultrashort laser pulses using chirped mirrors and glass wedges. *Opt. Express* **2012**, *20*, 688–697.

28. Miranda, M.; Arnold, C.L.; Fordell, T.; Silva, F.; Alonso, B.; Weigand, R.; L'Huillier, A.; Crespo, H. Characterization of broadband few-cycle laser pulses with the d-scan technique. *Opt. Express* **2012**, *20*, 18732–18743.

29. Loriot, V.; Gitzinger, G.; Forget, N. Self-referenced characterization of femtosecond laser pulses by chirp scan. *Opt. Express* **2013**, *21*, 24879–24893.

30. Sellmeier, W. Theory of anomalous light dispersion. *Ann. Phys. Chem.* **1871**, *143*, 271.

31. Malitson, I.H. Interspecimen comparison of refractive index of fused silica. *J. Opt. Soc. Am.* **1965**, *55*, 1205–1209.

32. Rasskazov, G.; Lozovoy, V.V.; Dantus, M. Spectral amplitude and phase noise characterization of titanium-sapphire lasers. *Opt. Express* **2015**, *23*, 23597–23602.

Chapter 3:
Spectroscopy

Sum-Frequency-Generation-Based Laser Sidebands for Tunable Femtosecond Raman Spectroscopy in the Ultraviolet

Liangdong Zhu, Weimin Liu, Yanli Wang and Chong Fang

Abstract: Femtosecond stimulated Raman spectroscopy (FSRS) is an emerging molecular structural dynamics technique for functional materials characterization typically in the visible to near-IR range. To expand its applications we have developed a versatile FSRS setup in the ultraviolet region. We use the combination of a narrowband, ~400 nm Raman pump from a home-built second harmonic bandwidth compressor and a tunable broadband probe pulse from sum-frequency-generation-based cascaded four-wave mixing (SFG-CFWM) laser sidebands in a thin BBO crystal. The ground state Raman spectrum of a laser dye Quinolon 390 in methanol that strongly absorbs at ~355 nm is systematically studied as a standard sample to provide previously unavailable spectroscopic characterization in the vibrational domain. Both the Stokes and anti-Stokes Raman spectra can be collected by selecting different orders of SFG-CFWM sidebands as the probe pulse. The stimulated Raman gain with the 402 nm Raman pump is >21 times larger than that with the 550 nm Raman pump when measured at the 1317 cm^{-1} peak for the aromatic ring deformation and ring-H rocking mode of the dye molecule, demonstrating that pre-resonance enhancement is effectively achieved in the unique UV-FSRS setup. This added tunability in the versatile and compact optical setup enables FSRS to better capture transient conformational snapshots of photosensitive molecules that absorb in the UV range.

Reprinted from *Appl. Sci.* Cite as: Zhu, L.; Liu, W.; Wang, Y.; Fang, C. Sum-Frequency-Generation-Based Laser Sidebands for Tunable Femtosecond Raman Spectroscopy in the Ultraviolet. *Appl. Sci.* **2015**, *5*, 48–61.

1. Introduction

The advent of femtosecond lasers has ushered in an exciting era of modern quantum chemistry and molecular spectroscopy [1,2] which has provided previously unavailable or hidden insights about structural dynamics, chemical reactivity, and biological functionality [2–7]. The ultrafast time duration of the incident laser pulses is key to dissect the electronic potential energy surface of the molecular system under investigation, and a time-delayed pump-probe setup is typically implemented to measure the system response on the intrinsic molecular timescale. In comparison to the widely used transient absorption technique that records the

175

electronic responses as a function of time, vibrational spectroscopy is intimately related to molecular structure and the associated normal modes, making the observed vibrational frequencies highly sensitive to the local environment of the chemical bond. By incorporating a preceding actinic pump pulse that induces photochemistry or other chemical reactions, vibrational transitions of the molecular system can be tracked in real time via IR absorption or Raman processes with ultrafast IR or visible laser pulses, typically on the femtosecond (fs) to picosecond (ps) timescale which can report on the incipient stage of photoinduced processes.

To exploit the full potential of ultrafast vibrational spectroscopy to characterize functional materials and biomolecules, we have developed femtosecond stimulated Raman spectroscopy (FSRS) as an emerging structural dynamics technique that has simultaneously high spectral and temporal resolutions [7–13]. The approach measures the ensemble average of system response so the observed structural evolution is insensitive to stochastic fluctuations but useful to report on functional atomic motions that are previously challenging to measure experimentally [2,7]. The conventional FSRS technique consists of an ~800 nm, ps Raman pump pulse from a grating-based spectral filter and a *ca.* 840–920 nm, fs Raman probe pulse from supercontinuum white light (SCWL) generation [9–11]. A preceding ~400 nm or 520–660 nm actinic pump pulse from second harmonic generation or noncollinear optical parametric amplification needs to be incorporated when the excited state molecular transformation is studied. One limiting factor for wider applications of FSRS is the wavelength tunability of incident pulses, particularly concerning the ps pump-fs probe pair that performs the stimulated Raman scattering process either in the electronic ground state (S_0) or excited state (e.g., S_1). Notably, different molecules have different potential energy landscapes so the vibrational energy levels and resonance Raman conditions vary greatly [14]. To turn FSRS into a more powerful and versatile spectroscopic toolset readily accessible to tackle a wide range of problems in energy and biology related fields, more technical innovations and optical advances are warranted.

In our earlier work, we reported the implementation of cascaded four-wave mixing (CFWM) in a thin transparent medium such as BK7 glass to generate broadband up-converted multicolor array (BUMA) signals [15]. One of these tunable, ultrabroad laser sidebands was used as the Raman probe in conjunction with an 800 nm Raman pump to collect the anti-Stokes Raman spectrum of a 1:1 v/v carbon tetrachloride:ethanol mixed standard solution [16] with high signal-to-noise ratio. We have also reported the BUMA sidebands in a 0.1-mm-thick BBO crystal at phase-matching condition for maximal second harmonic generation (SHG) [17]. The resultant SHG/sum-frequency-generation(SFG) assisted cascaded four-wave mixing processes lead to fs sideband signals from *ca.* 350–490 nm that are simultaneously enhanced due to $\chi^{(2)}$ and $\chi^{(3)}$-based four-wave optical parametric amplification [18].

Can this expanded versatility and tunability help expand the available optical methods for fs Raman probe generation? Can we investigate sample systems that primarily absorb in the UV, such as DNA/RNA molecules, metal-organic complexes that undergo ligand-metal charge transfer upon photoexcitation, and the functionally relevant amino acids (e.g., Trp and Tyr) in proteins and enzymes? Notably, the near-UV probe pulse can work well in conjunction with a ps Raman pump pulse, which is available from a second harmonic bandwidth compressor (SHBC) using the commercially available femtosecond 800 nm laser source [19]. In addition, the UV photoexcitation pulse at 267 nm can be readily generated from third harmonic generation of the 800 nm fundamental pulse.

In this article, we build on our previous results and report the construction and characterization of a versatile UV-FSRS setup incorporating both home-built second harmonic bandwidth compressor and SFG-based cascaded four-wave mixing. We demonstrate the feasibility of this setup in capturing the ground-state FSRS spectra of a laser dye Quinolon 390 (7-Dimethylamino-1-methyl-4-methoxy-8-azaquinolone-2, $C_{12}H_{15}N_3O_2$; Exciton Catalog No. 03900, or LD390) that has an absorption/emission peak at 355/390 nm in methanol. By tuning the ps Raman pump wavelength from visible (e.g., 550, 487 nm) to UV (e.g., 402 nm) and fs Raman probe wavelength in tandem based on supercontinuum white light generation as well as the broadband up-converted multicolor array technology, we achieve the pre-resonance enhancement factor of >21 for the stimulated Raman modes over a wide spectral window of >1400 cm^{-1}. These new results showcase the utility of tunable BUMA laser pulses in advancing the emerging FSRS technique, broadening its application potential to expose equilibrium and transient vibrational signatures of a wider array of photosensitive molecular systems that absorb in the UV to near-IR range [10,12].

2. Experimental Section

Our main optical setup to achieve tunable FSRS in the UV range uses a portion of the fundamental pulse (FP) output from a Ti:sapphire-based fs laser regenerative amplifier (Legend Elite-USP-1K-HE, Coherent, Inc.) seeded by a mode-locked Ti:sapphire oscillator (Mantis-5, Coherent). The FP of ~1 W at 800 nm center wavelength with 35 fs time duration (full-width-half-maximum, or fwhm) and 1 kHz repetition rate is split into two parts with a 9:1 ratio to pump the home-built second harmonic bandwidth compressor and the broadband up-converted multicolor array setup, respectively (see Figure 1). In the second harmonic bandwidth compressor section, the input beam is separated evenly into two arms, which go through reflective grating and cylindrical lens pairs and are stretched from fs to ps pulses with opposite chirps tuned to have the same magnitude [20]. After recombining the two arms at a 1-mm-thick Type-I BBO crystal ($\theta = 29.2°$) we achieve the chirp-free narrowband ps pulse centered at ~402 nm as a result of the chirp elimination effect [21]. The time duration of the second harmonic pulse is characterized by the optical Kerr effect (OKE) measurement with another 400 nm pulse from second harmonic generation of the FP followed by prism compression (40 fs, acting as the gate pulse), yielding a temporal profile with ~1.45 ps fwhm (Figure 2). To obtain the spectral width, we disperse the pulse with a 1200 grooves/mm, 500 nm blaze ruled reflective grating and image onto a CCD camera. After wavelength calibration with a mercury argon source (HG-1, Ocean Optics) across the UV/Vis range, the fwhm of the picosecond 402 nm pulse is measured to be ~11 cm^{-1}. This represents a time-bandwidth product of ~15.9 $ps\cdot cm^{-1}$ that is close to the Fourier-transform limit of a Gaussian-profile pulse (~14.7 $ps\cdot cm^{-1}$), indicating that the 402 nm pulse is largely chirp-free and can be used as the narrowband Raman pump pulse in FSRS.

In the broadband up-converted multicolor array setup, an FP and an SCWL (alternatively referred to as WL) pulse are loosely focused onto a 0.1-mm-thick BBO crystal to generate multiple sidebands via SFG-based cascaded four-wave mixing [17]. The crossing angle between the two incident fs pulses is ~6° to achieve balance between conversion efficiency and spatial separation of nascent sidebands [15,22,23]. The phase-matching condition of the BBO crystal is set to favor SFG. The first sideband on either the FP (S_{+1}) or WL (S_{-1}) side is selected with an iris diaphragm and used as the Raman probe (Figure 1). The pulse-to-pulse intensity stability is within 5%, which can be effectively averaged out by repeatedly collecting FSRS signals over several minutes (see below). Notably, the highly nonlinear pulse generation does not incur substantial intensity noise because we use low pump power to generate laser sidebands in the background-free directions (see Figure 1) so the interference effect with fundamental incident pulses is much reduced. Furthermore, the bandwidth of S_{+1} and S_{-1} is measured to be *ca.* 1400 and 1700 cm^{-1}, respectively, which supports the self-compression of these CFWM-induced sidebands to fs pulses [15,17]. To potentially achieve transform-limited pulses, further compression with accurate chirp compensation is needed [24,25].

The selected BUMA sideband and the SHBC output pulse are focused onto a 1-mm-thick quartz sample cell by an f = 12 cm off-axis parabolic mirror to avoid introducing additional chirps (e.g., if we use a focusing lens instead). Both incident beams pass through the sample solution containing 15 mM LD390 laser dye in methanol (Figure 1). The Raman pump is then blocked while the Raman probe pulse carrying the stimulated Raman scattering signal is re-collimated and focused into the spectrograph with a 1200 grooves/mm, 500 nm blaze ruled reflective grating. The dispersed signal is collected by a CCD array camera (Princeton Instruments, PIXIS 100F) that is synchronized with the laser at 1 kHz repetition rate to achieve shot-to-shot spectral acquisition. A phase-stable optical chopper (Newport 3501) in the Raman pump beampath at 500 Hz (also synchronized with the laser) ensures that one Raman spectrum can be collected within 2 ms through dividing the Raman probe profile with "Raman pump on" by "Raman pump off" (see Equation (1)).

Therefore, the recorded FSRS signal strength is typically expressed in the stimulated Raman gain:

$$\text{Raman Gain} = \text{Probe\_spectrum}_{\text{pump-on}} / \text{Probe\_spectrum}_{\text{pump-off}} - 1 \quad (1)$$

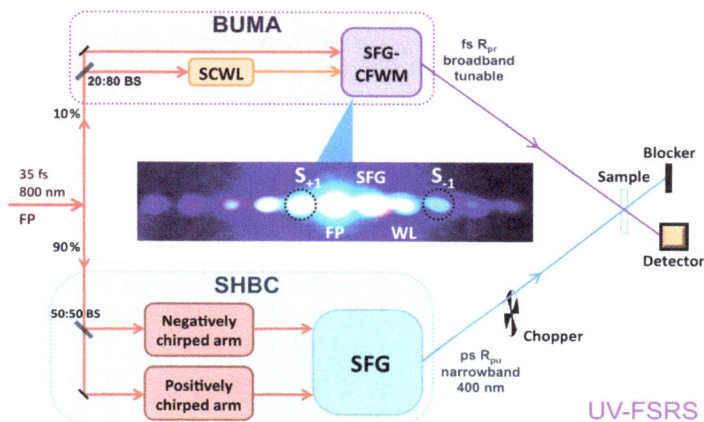

Figure 1. Schematic of the UV-Femtosecond stimulated Raman spectroscopy (FSRS) experimental setup. The fundamental laser output is split to separately pump a home-built second harmonic bandwidth compressor (SHBC, light blue shaded area) and a broadband up-converted multicolor array (BUMA, light violet shaded area) based on the unique sum-frequency-generation-based (SFG-based) cascaded four-wave mixing (CFWM) in a thin BBO crystal. A photograph of SFG-CFWM sideband signals on a sheet of white paper is shown with the first sideband on either side of SFG highlighted by black dotted circles. The first sideband either on the FP side (S_{+1}) or on the WL side (S_{-1}) is used as the Raman probe pulse in conjunction with the narrowband SHBC output as the Raman pump to record anti-Stokes and Stokes stimulated Raman spectrum, respectively. The 20:80 BS represents 20% Reflection and 80% Transmission.

We routinely collect the Raman spectrum with 3000 laser shots per point and 100 sets, so 150,000 Raman spectra are averaged to yield the final Raman spectrum as shown in Figures 3a and 4a with much improved signal-to-noise ratio. Experimentally we do not observe sharp noises that affect the detection sensitivity of the system, and the highly efficient data averaging within ~3 s for each recorded data trace largely removes the broad baseline fluctuations of the probe pulse (e.g., mostly up and down in intensity profile, not left and right along the frequency axis). All the experiments are performed at room temperature (21.9 °C) and ambient pressure (1 atm). To investigate the resonance Raman enhancement effect, we use the previously developed tunable FSRS in the visible to generate the ps Raman pump pulse (at 487 and 550 nm) in conjunction with an fs Raman probe pulse (to the red side of the pump) based on supercontinuum white light generation in a 2-mm-thick Z-cut sapphire plate followed by prism compression. Figure 3a displays the detailed comparison between ground-state FSRS spectra collected at various Raman pump-probe wavelengths, wherein the intensity noise level does not increase significantly as Raman pump wavelength approaches the electronic absorption peak. Figure 3b shows the computed Raman modes from density functional theory (DFT) B3LYP calculations in the electronic ground state using 6-311G+(d, p) basis sets for LD390 in methanol solution and the integral equation formalism polarizable continuum model (IEFPCM-methanol), performed by the *Gaussian 09* program [26].

3. Results and Discussion

The FSRS technology has been successfully applied to a number of important photosensitive molecular systems including rhodopsin [27], bacteriorhodopsin [28], phytochrome [29], organic dyes in solar cells [30], Fe(II) spin crossover in solution [31], fluorescent proteins [7,32,33], and calcium-ion-sensing protein biosensors [34–36]. The main goal of this work is to construct a versatile, tunable FSRS setup that extends the wavelength detection window to the UV regime with desired resonance Raman enhancement. As a result, a wider range of photochemical reaction pathways can be elucidated particularly for metal-organic complexes in solution (absorption peak below 300 nm) and tyrosine residues in proteins (max absorption at ~276 nm) in conjunction with a femtosecond actinic pump pulse. Because FSRS is a stimulated Raman technique, the concomitant generation of a pair of ps-Raman-pump and fs-Raman-probe pulses is required.

3.1. UV-FSRS Setup with SHBC and SFG-CFWM

Starting from the fs 800 nm laser amplifier system, we choose to exploit a home-built single-grating-based second harmonic bandwidth compressor to produce a ~400 nm, ps pulse [20] as the Raman pump. To conveniently generate an accompanying Raman probe, we rely on the SFG-CFWM method that can be readily tuned by varying the time delay between the two incident pulses or selecting a different sideband on either side of the FP beam (see Figure 1, middle). The wavelength tunability of those sidebands has been discussed in our previous reports [15,17]. Figure 2a shows temporal characterization of the second harmonic bandwidth compressor output at 402 nm with ~1.45 ps pulse duration (fwhm). Figure 2b displays the relative spectral position of the narrowband Raman pump and two distinct femtosecond BUMA sidebands from SFG-CFWM processes in a thin BBO crystal, S_{+1} and S_{-1}, which enable the collection of anti-Stokes and Stokes Raman spectrum, respectively. Furthermore, we have demonstrated the SFG-CFWM sidebands spanning a broad UV to visible spectral range from *ca.* 350–490 nm [17], which can be potentially pushed toward shorter wavelengths upon increasing the pump power and/or reducing the incident beam crossing angle [23,24].

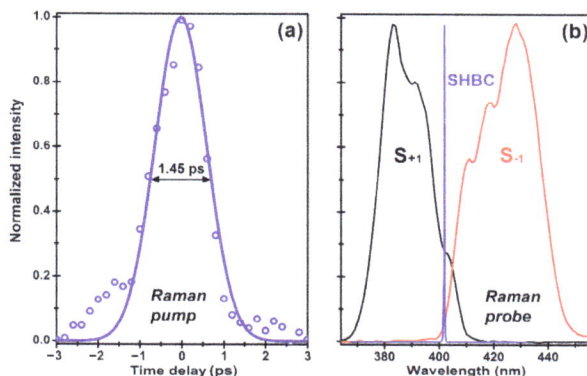

Figure 2. Spectral characterization of the Raman pump and probe pulse pair for FSRS. (**a**) Temporal profile of the narrowband Raman pump measured from optical Kerr effect (OKE). The pulse duration of ~1.45 ps is obtained from the fwhm of the Gaussian fit (blue solid curve) to the time-resolved experimental data points (blue open circles). (**b**) Spectra of the SHBC output as the Raman pump in (**a**) (blue) and the first two BUMA sidebands as the Raman probe, S_{+1} on the FP side (black) and S_{-1} on the WL side (red), respectively.

3.2. Ground-State FSRS of Laser Dye LD390

To demonstrate the feasibility of the aforementioned UV-FSRS setup, we select LD390 as the molecular sample system because this laser dye in methanol strongly absorbs 355 nm light, while the solvent only has two major Raman peaks at ~1033 and 1460 cm^{-1}. To our best knowledge, the standard or spontaneous Raman spectrum of LD390 has not been reported, so the measurement here represents a new spectroscopic characterization of this commercial laser dye molecule and its vibrational motions in solution. After equal amount of solution and solvent data collection, average and subtraction which remove most of the systematic noise and laser fluctuation effect, the pure ground-state Raman spectrum of LD390 is shown in Figure 3a that has a number of prominent peaks between *ca.* 300–1700 cm^{-1}. These are Stokes Raman spectra because the probe pulse is S_{-1} on the WL side (Figure 1) and to the red of the pump pulse (Figure 2b). Based on the UV/Vis spectrum in Figure 3a insert, the 402 nm Raman pump represents the closest frequency position to the electronic absorption peak (*i.e.*, 355 nm) among the three Raman pump wavelengths being used, and the pre-resonance enhancement factor reaches >21 at the 1317/1348 cm^{-1} peak doublet while all the other experimental conditions are unchanged. The observed peak intensity decreases if the Raman pump wavelength is tuned away from the electronic absorption peak position. Notably, minimal interference below 400 cm^{-1} makes it feasible to study lower frequency regime of the Raman spectrum. This likely arises from good solubility of the dye molecule in methanol and less Raman pump scattering into the Raman probe beampath (*i.e.*, FSRS signal direction) [11,13].

Table 1. Ground-state FSRS vibrational peak frequencies and mode assignments aided by calculations.

S_0 calc. [a] (cm^{-1})	S_0 FSRS [b] (cm^{-1})	Vibrational mode assignment [c]
665	663	Ring in-plane asymmetric deformation
711	714	Ring asymmetric breathing with N1–CH$_3$ stretching
1068	1069	A-ring deformation and H rocking, B-ring small-scale breathing, and N13–(CH$_3$)$_2$ H twisting
1312	1317	N1–C2 stretching with ring asymmetric deformation and ring-H rocking, and (N1)–CH$_3$ methyl group bending
1357	1348	C7–N8 stretching and A-ring H rocking, A-ring in-plane deformation with some C9–C10 stretching
1393	1390	N1–CH$_3$ stretching and methyl group symmetric bending, A-ring in-plane deformation, and ring-H rocking
1613	1615	Ring C=C and C=N stretching, ring-H rocking with C2=O11 stretching

[a] Vibrational normal mode frequencies are obtained from DFT B3LYP calculations in S_0 using 6-311G+(d, p) basis sets for LD390 in methanol with *Gaussian 09* program [26]. The scaling factor of 0.99 is used to compare the calculated frequency with experimental result. [b] The experimentally observed frequencies of the ground-state Raman peaks of 15 mM LD390 in pure methanol using tunable FSRS technology in the UV to visible range. [c] Vibrational motions are assigned based on DFT calculation results. Only major vibrational modes are listed with the atomic numbering defined in Figure 3b insert.

Since the standard Raman spectrum of LD390 is not readily available from literature and it is useful to correlate observed peaks to characteristic nuclear motions, we perform electronic ground-state DFT calculations in *Gaussian* program [26] to facilitate vibrational normal mode assignment. The overall match between the experimental and calculated spectrum is very good with a frequency scaling factor of 0.99 [37]. The major vibrational modes are listed in Table 1. Notably, the correspondence between the calculated Raman spectrum and the measured one is not exact (Figure 3). This is understandable because the Gaussian DFT calculation concerns an "unrestricted" single molecule in a polarizable continuum to model solvation effects (*i.e.*, we used IEFPCM-methanol, see above). In the real spectroscopic measurement, the ensemble average of solvated dye molecules LD390 in methanol solution is measured and the Raman mode polarizability is intimately determined by the extensive hydrogen (H)-bonding network around the chromophore. For example, the calculated strong modes between *ca.* 1450–1600 cm^{-1} (see Figure 3b) become much weaker in the ground-state FSRS spectrum (Figure 3a), suggesting that the corresponding mode polarizability decreases significantly and/or mode frequency shifts as a result of H-bonding matrix. It is also notable that the two strongest peaks observed at ~1317, 1348 cm^{-1} both consist of C–N stretching and ring-H rocking motions plus ring in-plane deformations on both aromatic rings of the dye molecule. The large change in conjugation and electronic polarizability over the two-ring system leads to the observed strong Raman gain in comparison to other vibrational modes. Moreover, the amplification of the Stokes Raman spectrum primarily applies to the solute signal but not to spectral noise, so the experimental signal-to-noise ratio is greatly enhanced with the ~400 nm Raman pump and should be beneficial to characterize functional materials and molecular systems with intrinsically small electric polarizabilities [20,38].

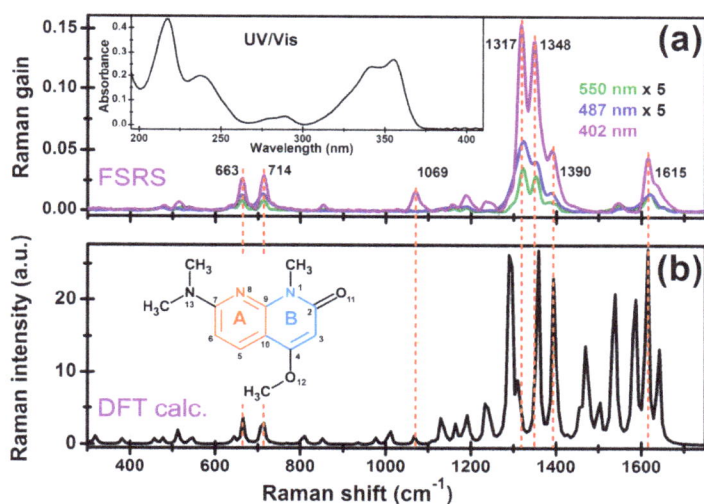

Figure 3. Ground-state Stokes FSRS of LD390 in methanol. (**a**) Experimental stimulated Raman spectra in S$_0$ with the Raman pump at 550 nm (green), 487 nm (blue), and 402 nm (violet) and Raman probe to its red side, respectively. The former two spectral traces are enlarged by 5 times for direct comparison with the spectrum collected with 402 nm pump. Prominent vibrational peaks are marked with frequencies labeled in black. The UV/Vis electronic absorption spectrum is shown in the insert. (**b**) Density functional theory (DFT)-based *Gaussian* calculated spectrum of LD390 in methanol with a uniform peak width of 8 cm^{-1} (*i.e.*, default fwhm in the program). The molecular structure of the dye is depicted in the insert with two aromatic rings labeled in A (orange) and B (cyan). The key atomic sites are numbered from 1–13.

3.3. Comparison between the Stokes and Anti-Stokes FSRS

It is notable that both the Stokes and anti-Stokes Raman spectra can be conveniently captured by FSRS gain/loss measurement depending upon the relative wavelengths of the Raman pump and probe pulses, while the latter can be switched between various BUMA sidebands (e.g., shifting the pinhole position) or tuned within the same sideband (e.g., varying the time delay between FP and WL) with ease (see Figure 1). Based on partition functions in thermodynamics, there is less population on the first excited vibrational state (*i.e.*, quantum number $v = 1$) than that on the ground state ($v = 0$). At room temperature the thermal energy $1 k_B T$ amounts to ~200 cm^{-1} so all the vibrational modes (e.g., >300 cm^{-1} in Figure 3a) should display weaker anti-Stokes spontaneous Raman peaks than the corresponding Stokes peaks particularly for high-frequency modes. In contrast, FSRS signal strength is normalized by the probe intensity (see Equation (1)) but is typically proportional to Raman pump power [9,39] and to the square of the SRS nonlinear coefficient for either the Stokes or anti-Stokes signal [40,41]. Figure 4a shows that the ground-state FSRS anti-Stokes Raman spectrum we collected using S_{+1} on the FP side as the probe (*i.e.*, to the blue of the 402 nm pump pulse in Figure 2b) is much stronger than the Stokes spectrum. This unusual, opposite trend indicates that some other factors contribute to the Raman signal strength beyond the Raman pump power and third-order nonlinear polarizabilities [41]. Can it arise from resonance enhancement because this is a stimulated Raman technique [14,42]? If so, how does the Raman pump wavelength compare to the 0–0 vertical transition energy between the electronic ground state and excited state of LD390 in both FSRS measurements?

We list all the anti-Stokes over Stokes peak intensity ratios in Figure 4a insert and it becomes apparent that the two modes below 750 cm^{-1} have a ratio below 1.8 while the modes above 1000 cm^{-1} all have a ratio above 3.0. The overall trend is that the Raman peak gets stronger as the vibrational frequency increases. Given that the anti-Stokes process originates from the higher-lying vibrational state (e.g., $v = 1$) and terminates at the lower-lying vibrational state (e.g., $v = 0$), this experimental trend can be explained by the principle of resonance Raman enhancement because the 402 nm pump pulse being used still falls short of the LD390 electronic absorption peak of ~ 355 nm (Figure 4b). As a result, for the anti-Stokes transition, the 663 (1615) cm^{-1} modes correspond to an "effective" Raman pump wavelength of 392 (377) nm, making the latter Raman mode much stronger because a 377 nm pump is in closer proximity to the 355 nm electronic gap than a 392 nm pump. This reasoning is further corroborated by experimental data in Figure 3a, and paves the way to enhance higher-frequency Raman modes regardless of their intrinsic electronic polarizability. For the Stokes spectrum that starts from the ground state and ends on the $v = 1$ state in S_0, the vibrational transition frequency does not affect the energy relation between the 402 nm Raman pump and the 355 nm electronic energy gap (*i.e.*, no addition of the vibrational frequency can occur to bring the 402 nm Raman pump closer to the solute S_1 state, see Figure 4b), hence the intensity ratio between Raman peaks is mostly determined by the mode-dependent polarizability [13,43].

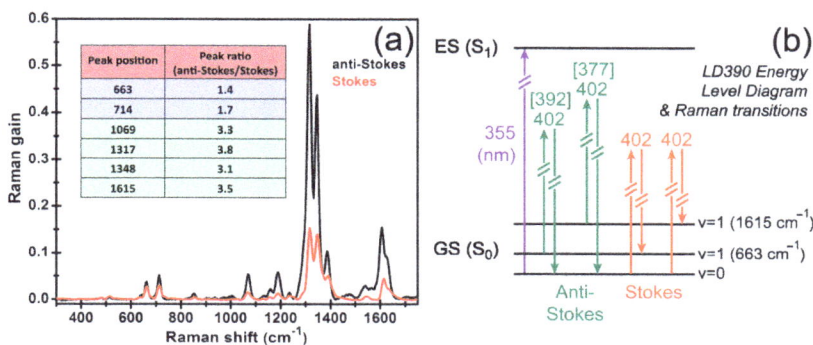

Figure 4. (a) Comparison between the Stokes (red) and anti-Stokes (black) ground-state FSRS data for LD390 in methanol solution. The first-order UV-BUMA sideband S_{+1}/S_{-1} on the fundamental pulse (FP)/white light (WL) side acts as the Raman probe for anti-Stokes/Stokes FSRS with the 402 nm Raman pump pulse, respectively. The frequency axes are calibrated and for direct spectral comparison, the anti-Stokes Raman shift axis as well as the Raman peak intensities are multiplied by –1. The insert tabulates the observed peak intensity ratios of several major vibrational modes between 600–1700 cm^{-1}. (b) The spectroscopic origin of the observed Raman intensity ratios can be understood by the molecular energy level diagram of LD390. Two characteristic vibrational modes are shown (vibrational quantum number v = 1) with the relative energy differences between various Raman transition configurations depicted by colored vertical arrowed lines. The numbers in brackets represent effective Raman pump wavelengths in nm unit (see Section 3.3.).

4. Conclusions

In summary, we have developed a unique UV-FSRS setup with a home-built second harmonic bandwidth compressor output (~400 nm center wavelength, 1.5 ps fwhm) as the Raman pump and various broadband up-converted multicolor array (BUMA) sideband laser pulses (*ca.* 360—460 nm, fs) as the Raman probe to obtain the stimulated Raman spectrum. The BUMA signals in this work arise from SFG/SHG-based cascaded four-wave mixing processes in a thin BBO crystal. Two other Raman pump wavelengths are achieved using a tunable FSRS setup in the visible range and the resultant Stokes spectrum of 15 mM LD390 in methanol is >21 times weaker due to larger mismatch between the Raman pump wavelength and the electronic absorption peak frequency of the laser dye. This manifests the advantage of using tunable ps pulses to study molecules with different absorption profiles over a wide spectral range. The Raman spectrum of LD390 is collected over a ~1400 cm^{-1} detection window for the first time with vibrational mode assignments aided by *Gaussian* DFT calculations. Using a ~400 nm ps Raman pump, the anti-Stokes Raman spectrum turns out to be much stronger than the Stokes spectrum mainly due to pre-resonance enhancement involving the vibrational energy gap in S_0, which is confirmed by the relative intensity ratio change between the low- and high-frequency vibrational modes.

The versatile and compact approach of generating tunable probe pulses in the UV should make the FSRS technology more accessible to many laboratories for elucidation of molecular conformational dynamics in the electronic ground state, as well as excited state upon incorporation of a preceding fs photoexcitation pulse [10,11,13]. This methodology also paves the way to harness the broadband tunability of multi-color laser sidebands to study molecules that primarily absorb in the UV which include metal-organic complexes such as triphenylbismuth in methanol solution and biomolecules such as DNA and tyrosine derivatives in water. Related studies are currently underway.

Acknowledgments: This project is supported by the Oregon State University Faculty Startup Research Grant and Research Equipment Reserve Fund (to C.F.) and the National Science Foundation (NSF) CAREER award (grant number CHE-1455353, to C.F. since Feb. 2015). We thank Joseph Nibler for the laser dye sample. We also acknowledge graduate research assistantship (to L.Z.) from the NSF Center for Sustainable Materials Chemistry (grant number CHE-1102637).

Author Contributions: C.F. designed and supervised research; L.Z., W.L., and C.F. contributed new analytic tools; L.Z. and W.L. performed research; L.Z., Y.W., and C.F. analyzed data; and C.F. wrote the paper with discussions from all the authors.

Conflicts of Interest: The authors declare no conflict of interest.

References

1.   Zewail, A.H. *Femtochemistry: Ultrafast Dynamics of the Chemical Bond*; World Scientific: Singapore, 1994.

2.   Hochstrasser, R.M. Two-dimensional spectroscopy at infrared and optical frequencies. *Proc. Natl. Acad. Sci. USA* **2007**, *104*, 14190–14196.

3.   Zheng, J.; Kwak, K.; Asbury, J.; Chen, X.; Piletic, I.R.; Fayer, M.D. Ultrafast dynamics of solute-solvent complexation observed at thermal equilibrium in real time. *Science* **2005**, *309*, 1338–1343.

4.   Fang, C.; Senes, A.; Cristian, L.; DeGrado, W.F.; Hochstrasser, R.M. Amide vibrations are delocalized across the hydrophobic interface of a transmembrane helix dimer. *Proc. Natl. Acad. Sci. USA* **2006**, *103*, 16740–16745.

5. Engel, G.S.; Calhoun, T.R.; Read, E.L.; Ahn, T.-K.; Mancal, T.; Cheng, Y.-C.; Blankenship, R.E.; Fleming, G.R. Evidence for wavelike energy transfer through quantum coherence in photosynthetic systems. *Nature* **2007**, *446*, 782–786.

6. Fang, C.; Bauman, J.D.; Das, K.; Remorino, A.; Arnold, E.; Hochstrasser, R.M. Two-dimensional infrared spectra reveal relaxation of the nonnucleoside inhibitor TMC278 complexed with the HIV-1 reverse transcriptase. *Proc. Natl. Acad. Sci. USA* **2008**, *105*, 1472–1477.

7. Fang, C.; Frontiera, R.R.; Tran, R.; Mathies, R.A. Mapping GFP structure evolution during proton transfer with femtosecond Raman spectroscopy. *Nature* **2009**, *462*, 200–204.

8. Yoshizawa, M.; Kurosawa, M. Femtosecond time-resolved Raman spectroscopy using stimulated Raman scattering. *Phys. Rev. A* **1999**, *61*, 013808.

9. McCamant, D.W.; Kukura, P.; Yoon, S.; Mathies, R.A. Femtosecond broadband stimulated Raman spectroscopy: Apparatus and methods. *Rev. Sci. Instrum.* **2004**, *75*, 4971–4980.

10. Frontiera, R.R.; Fang, C.; Dasgupta, J.; Mathies, R.A. Probing structural evolution along multidimensional reaction coordinates with femtosecond stimulated Raman spectroscopy. *Phys. Chem. Chem. Phys.* **2012**, *14*, 405–414.

11. Liu, W.; Han, F.; Smith, C.; Fang, C. Ultrafast conformational dynamics of pyranine during excited state proton transfer in aqueous solution revealed by femtosecond stimulated Raman spectroscopy. *J. Phys. Chem. B* **2012**, *116*, 10535–10550.

12. Dasgupta, J.; Frontiera, R.R.; Fang, C.; Mathies, R.A. Femtosecond stimulated Raman spectroscopy. In *Encyclopedia of Biophysics*; Roberts, G.C.K., Ed.; Springer: Berlin, Germany, 2013; pp. 745–759.

13. Han, F.; Liu, W.; Fang, C. Excited-state proton transfer of photoexcited pyranine in water observed by femtosecond stimulated Raman spectroscopy. *Chem. Phys.* **2013**, *422*, 204–219.

14. Myers, A.B.; Mathies, R.A. Resonance Raman intensities: A probe of excited-state structure and dynamics. In *Biological Applications of Raman Spectroscopy*; Spiro, T.G., Ed.; John Wiley & Sons, Inc.: New York, NY, USA, 1987; Volume 2, pp. 1–58.

15. Liu, W.; Zhu, L.; Wang, L.; Fang, C. Cascaded four-wave mixing for broadband tunable laser sideband generation. *Opt. Lett.* **2013**, *38*, 1772–1774.

16. Zhu, L.; Liu, W.; Fang, C. Tunable sideband laser from cascaded four-wave mixing in thin glass for ultra-broadband femtosecond stimulated Raman spectroscopy. *Appl. Phys. Lett.* **2013**, *103*, 061110.

17. Liu, W.; Zhu, L.; Fang, C. Observation of sum-frequency-generation-induced cascaded four-wave mixing using two crossing femtosecond laser pulses in a 0.1 mm beta-barium-borate crystal. *Opt. Lett.* **2012**, *37*, 3783–3785.

18. Zhu, L.; Liu, W.; Wang, L.; Fang, C. Parametric amplification-assisted cascaded four-wave mixing for ultrabroad laser sideband generation in a thin transparent medium. *Laser Phys. Lett.* **2014**, *11*, 075301.

19. Laimgruber, S.; Schachenmayr, H.; Schmidt, B.; Zinth, W.; Gilch, P. A femtosecond stimulated Raman spectrograph for the near ultraviolet. *Appl. Phys. B* **2006**, *85*, 557–564.

20. Zhu, L.; Liu, W.; Fang, C. A versatile femtosecond stimulated Raman spectroscopy setup with tunable pulses in the visible to near infrared. *Appl. Phys. Lett.* **2014**, *105*, 041106.

21. Raoult, F.; Boscheron, A.C.L.; Husson, D.; Sauteret, C.; Modena, A.; Malka, V.; Dorchies, F.; Migus, A. Efficient generation of narrow-bandwidth picosecond pulses by frequency doubling of femtosecond chirped pulses. *Opt. Lett.* **1998**, *23*, 1117–1119.

22. Crespo, H.; Mendonca, J.T.; Dos Santos, A. Cascaded highly nondegenerate four-wave-mixing phenomenon in transparent isotropic condensed media. *Opt. Lett.* **2000**, *25*, 829–831.

23. Liu, J.; Kobayashi, T. Cascaded four-wave mixing and multicolored arrays generation in a sapphire plate by using two crossing beams of femtosecond laser. *Opt. Express* **2008**, *16*, 22119–22125.

24. Weigand, R.; Mendonca, J.T.; Crespo, H.M. Cascaded nondegenerate four-wave-mixing technique for high-power single-cycle pulse synthesis in the visible and ultraviolet ranges. *Phys. Rev. A* **2009**, *79*, 063838.

25. Shitamichi, O.; Kida, Y.; Imasaka, T. Chirped-pulse four-wave Raman mixing in molecular hydrogen. *Appl. Phys. B* **2014**, *117*, 723–730.

26. Frisch, M.J.; Trucks, G.W.; Schlegel, H.B.; Scuseria, G.E.; Robb, M.A.; Cheeseman, J.R.; Scalmani, G.; Barone, V.; Mennucci, B.; Petersson, G.A.; *et al. Gaussian 09*; Revision B.1; Gaussian, Inc.: Wallingford, CT, USA, 2009.

27. Kukura, P.; McCamant, D.W.; Yoon, S.; Wandschneider, D.B.; Mathies, R.A. Structural observation of the primary isomerization in vision with femtosecond-stimulated Raman. *Science* **2005**, *310*, 1006–1009.

28. Shim, S.; Dasgupta, J.; Mathies, R.A. Femtosecond time-resolved stimulated Raman reveals the birth of bacteriorhodopsin's J and K intermediates. *J. Am. Chem. Soc.* **2009**, *131*, 7592–7597.

29. Dasgupta, J.; Frontiera, R.R.; Taylor, K.C.; Lagarias, J.C.; Mathies, R.A. Ultrafast excited-state isomerization in phytochrome revealed by femtosecond stimulated Raman spectroscopy. *Proc. Natl. Acad. Sci. USA* **2009**, *106*, 1784–1789.

30. Frontiera, R.R.; Dasgupta, J.; Mathies, R.A. Probing interfacial electron transfer in Coumarin 343 sensitized TiO_2 nanoparticles with femtosecond stimulated Raman. *J. Am. Chem. Soc.* **2009**, *131*, 15630–15632.

31. Smeigh, A.L.; Creelman, M.; Mathies, R.A.; McCusker, J.K. Femtosecond time-resolved optical and Raman spectroscopy of photoinduced spin crossover: Temporal resolution of low-to-high spin optical switching. *J. Am. Chem. Soc.* **2008**, *130*, 14105–14107.

32. Kuramochi, H.; Takeuchi, S.; Tahara, T. Ultrafast structural evolution of photoactive yellow protein chromophore revealed by ultraviolet resonance femtosecond stimulated Raman spectroscopy. *J. Phys. Chem. Lett.* **2012**, *3*, 2025–2029.

33. Creelman, M.; Kumauchi, M.; Hoff, W.D.; Mathies, R.A. Chromophore dynamics in the PYP photocycle from femtosecond stimulated Raman spectroscopy. *J. Phys. Chem. B* **2014**, *118*, 659–667.

34. Oscar, B.G.; Liu, W.; Zhao, Y.; Tang, L.; Wang, Y.; Campbell, R.E.; Fang, C. Excited-state structural dynamics of a dual-emission calmodulin-green fluorescent protein sensor for calcium ion imaging. *Proc. Natl. Acad. Sci. USA* **2014**, *111*, 10191–10196.

35. Tang, L.; Liu, W.; Wang, Y.; Zhao, Y.; Oscar, B.G.; Campbell, R.E.; Fang, C. Unraveling ultrafast photoinduced proton transfer dynamics in a fluorescent protein biosensor for Ca^{2+} imaging. *Chem. Eur. J.* **2015**, *21*, 6481–6490.

36. Wang, Y.; Tang, L.; Liu, W.; Zhao, Y.; Oscar, B.G.; Campbell, R.E.; Fang, C. Excited state structural events of a dual-emission fluorescent protein biosensor for Ca^{2+} imaging studied by femtosecond stimulated Raman spectroscopy. *J. Phys. Chem. B* **2015**, *119*, 2204–2218.

37. Tozzini, V.; Nifosì, R. Ab initio molecular dynamics of the green fluorescent protein (GFP) chromophore: An insight into the photoinduced dynamics of green fluorescent proteins. *J. Phys. Chem. B* **2001**, *105*, 5797–5803.

38. Wang, W.; Liu, W.; Chang, I.-Y.; Wills, L.A.; Zakharov, L.N.; Boettcher, S.W.; Cheong, P.H.-Y.; Fang, C.; Keszler, D.A. Electrolytic synthesis of aqueous aluminum nanoclusters and *in situ* characterization by femtosecond Raman spectroscopy & computations. *Proc. Natl. Acad. Sci. USA* **2013**, *110*, 18397–18401.

39. Frontiera, R.R.; Shim, S.; Mathies, R.A. Origin of negative and dispersive features in anti-Stokes and resonance femtosecond stimulated Raman spectroscopy. *J. Chem. Phys.* **2008**, *129*, 064507.

40. Lee, S.Y.; Heller, E.J. Time-dependent theory of Raman scattering. *J. Chem. Phys.* **1979**, *71*, 4777–4788.

41. Lee, S.-Y.; Zhang, D.; McCamant, D.W.; Kukura, P.; Mathies, R.A. Theory of femtosecond stimulated Raman spectroscopy. *J. Chem. Phys.* **2004**, *121*, 3632–3642.

42. Shim, S.; Stuart, C.M.; Mathies, R.A. Resonance Raman cross-sections and vibronic analysis of Rhodamine 6G from broadband stimulated Raman spectroscopy. *ChemPhysChem* **2008**, *9*, 697–699.

43. McHale, J.L. *Molecular Spectroscopy*; Prentice-Hall: Upper Saddle River, NJ, USA, 1999.

MDPI AG

St. Alban-Anlage 66

4052 Basel, Switzerland

Tel. +41 61 683 77 34

Fax +41 61 302 89 18

http://www.mdpi.com

Applied Sciences Editorial Office

E-mail: applsci@mdpi.com

http://www.mdpi.com/journal/applsci